understanding chemical thermodynamics

GEORGE C. PIMENTEL
University of California, Berkeley

RICHARD D. SPRATLEY
University of British Columbia

understanding chemical thermodynamics

HOLDEN-DAY, INC.
San Francisco
Cambridge
London
Amsterdam

Library of Congress Catalog Card Number 69-13419

Printed in the United States of America

preface

Introductory chemistry courses have evolved rapidly in the last few years. The inclusion of unifying principles, in particular, chemical thermodynamics, is now a well-established pattern. This desirable development has raised pedagogic difficulties. Earlier, thermodynamics was presented at an advanced level at which mathematical competence could be assumed. This traditional approach, with an adjustment in level, has been retained in recently produced materials for accelerated university freshman chemistry courses. These monographs and texts are successful in the hands of specially prepared students who are taking calculus concurrently.

We address this monograph to a larger and, perhaps, a more important group—those who habit the general introductory chemistry course and who will not necessarily continue in the physical sciences. This group, too, needs to see the coherence of chemistry; the interplay between facts and theories. Thermodynamics provides this opportunity. This group's interest, however, will not be primarily in facile manipulation of thermodynamic equations, but rather in the meanings and implications of thermodynamic principles. That is the emphasis we wish to pursue.

Thermodynamics is the study of energy and randomness. In chemistry, the First Law of Thermodynamics tells us of the energy effects that accompany chemical reactions. The Second Law of Thermodynamics tells what reactions can actually occur. The latter—understanding the motivation behind chemical changes—is the more important accomplishment of chemical thermodynamics. It is also the more difficult to display, unobscured by mathematical apparatus and the abstraction of the entropy concept. So we have focussed attention on the meaning of the Second Law with a simple (but valid) development of its statistical mechanical foundation. We do so with the conviction that entropy is as easy to understand as energy, if the mathematical foliage is cleared away.

The material presented here is distilled from the authors' classroom experience at Berkeley over the past five years in the accelerated freshman chemistry course. It was used successfully during the 1967–68 academic year in general freshman chemistry courses at the University of California. Some chemical familiarity is assumed, so these topics are lodged near the middle of the academic year. This placement permits, in conclusion, a return to descriptive chemistry with an opportunity to exercise thermodynamic ideas and to illustrate their power.

We have the usual debts to our colleagues, whose encouragement and critiques were invaluable. Our students, too, have contributed more than they could know, through their conscientious efforts as our material was still taking form. Finally, we warmly thank Miss Susan Arbuckle for her tireless efforts, her excellent typing, her delicious candies, and her everlasting cheerful spirit.

February, 1969

George C. Pimentel
Richard D. Spratley

contents

understanding chemical thermodynamics

prologue

Heat has preoccupied man since his earliest conscious concern over survival. Countless men have shivered before their fires, entranced by the dancing flames, wondering how a dead tree limb could release this marvelous display with its life-sustaining warmth. Fire was given a prominent role in the alchemists' first rudimentary attempts to organize and apply experience with chemical
300 B.C. change. They perpetuated the Greeks' view that matter was composed of the four elements—earth, air, fire, and water—up until the sixteenth century!

During this 2000-year period, all of science remained in the bud, awaiting an intellectual Spring before it could blossom. The end of this long Winter required the sunlight of experimentation—observational knowledge must guide ideas, and laboratory challenge must test them. This favorable climate finally developed about 250 years ago and only then did **thermodynamics,** *the study of heat and its transformations,* come into being as a systematic branch of science.

Thermodynamics in physics

Caloric: One of the four elements

Aristotle, in 350 B.C., laid a stamp upon the pseudo-scientific thinking about heat that persisted for very many centuries. Air, he said, consisted of
350 B.C. caloric (heat) plus moisture, while fire was made of caloric plus dryness. Thus, heat was considered to be a substance until the turn of the nineteenth century when Sir Humphrey Davy confronted himself with an experiment on the frictional melting of ice by clockwork run in a vacuum.

"Heat, when produced by friction, cannot be
1800 collected from the bodies in contact, and it was proved by the . . . experiment that the increase in temperature consequent on friction cannot arise from diminution of capacity or from oxydation. . . . Since . . . it is produced in neither of these modes, it cannot be considered as matter. It has then been experimentally demonstrated that caloric, or the matter of heat, does not exist." *

This bold contradiction of prevailing dogma was published in 1799, coincident with the now better

*W. F. Magie, *A Source Book in Physics,* p. 164, Harvard University Press, Cambridge, 1963.

known, independent studies by Count Rumford on the heat produced during the boring of a cannon. With astonishing insight in a pre-atomic period, he said:

". . . anything which any *insulated* body . . . can continue to furnish *without limitation* cannot possibly be a *material substance:* and it appears to me to be extremely difficult, if not quite impossible, to form any distinct idea of anything, capable of being excited and communicated, in the manner the Heat was excited and communicated in these Experiments, except it be MOTION."*

Vis viva: A quantity conserved

During the eighteenth century, observations on moving bodies were incorporated with intuitive ideas of force and inertial mass into quantitative laws of motion. As these laws took form, they revealed an important aspect of simple collisional processes: the sum of the quantities $\frac{1}{2}mv^2$ for the colliding bodies had the same
1750 value after the collision as before the collision! Here was a quantity that remained constant during the collisional event—hence a quantity that provided a basis for prediction. That is the business of science:

—to order observations,
—to find guiding principles,
—to use these principles to develop expectations.

The constant quantity was given a name: *vis viva.* As its usefulness entrenched it in man's thinking, *vis viva* took on a mental tangibility. Instead of saying it was a "constant of motion," scientists began saying it was "conserved." The name *vis viva* was gradually superseded by its modern name, *energy.*

Heat and energy identified

Unfortunately, the elegance of celestial mechanics displayed nightly in the heavens above was tarnished by man's laboratory experience. His most accurate experiments persistently showed that *vis viva* was, in fact, not constant. The discrepancy seemed to lie in the frictional processes which were always accompanied by manifestations of heat. Although the logical basis for belief in *vis viva,* its constancy, was thereby negated, scientists found the conservation idea so reassuring that heat was accepted as a transformed kind of *vis viva.* In 1845, James Prescott Joule made the relationship quantitative by a series of experiments in which work was expended so as to produce a temperature rise. His celebrated paddle-wheel experiment was only one of his several studies:

"The paddle moved with great resistance in the can of water, so that the weights . . . descended at the slow rate of about one

* *Ibid.*, p. 161.

1850 foot per second . . . when the weights had descended . . . they had to be wound up again in order to renew the motion of the pulley. After this operation had been repeated sixteen times, the increase of the temperature of the water was ascertained by means of a very sensitive and accurate thermometer . . . it appeared that for each degree of heat evolved by the friction of water a mechanical power equal to that which can raise a weight of 890 lb to the height of one foot had been expended."*

And then, potential energy

Thus the *conservation of energy* became a working doctrine. The next developments can be foreseen. A weight is lifted onto a table. Work is expended, but no heat is in evidence. Now where is the energy? Never mind—we'll say it is "stored" in the weight. After all, experience tells us that the work can be regained any time we wish, merely by pushing the weight off the table. The elevated weight has the *potentiality* of returning to us the energy we have expended. Energy is still conserved because the weight possesses "potential energy." How much? Anything missing after we subtract any frictional heat produced from the work done!

How far we have come from the dynamic quality of *vis viva*, in which the constancy of properties of moving bodies—the sum of $\frac{1}{2}mv^2$—was the excuse for the concept. Now a dormant weight on a table possesses energy—in amount exactly equated to the *lack* of constancy of the *vis viva* after heat is taken into account.

And now—all is energy

And so the energy concept has become more and more all-embracing. Now we have energy of translation, rotation, vibration, chemical energy, magnetic energy, electrostatic energy, nuclear energy, gravitational energy—even nervous energy.

When, in beta decay of the nucleus, the ejected electrons indicated that they did not possess all of the available energy, physicists readily accepted the postulate of the neutrino, an invisible particle having no charge and no detectable mass, to carry away the remaining energy and save the conservation of energy (after all, it is 1933 and conservation of energy is now a Law!). Could the present confusion about the myriad of nuclear particles now catalogued be rooted in our insistence on this (or another) Conservation Law?

1911 But the grand sweep is yet to be mentioned. Einstein, in 1911, reconsidered the significance of relative motion when the Michelson–Morley experiment failed to reveal the existence of an aether. He concluded that mass must also be a manifestation of energy—$E = mc^2$! Now everything is energy! We have come full circle. Aristotle's mental meanderings were correct: caloric *is* substance

* *Ibid.*, p. 206.

after all! And air *is* heat and moisture! Now the energy concept is so all-inclusive that it hardly seems interesting to say "energy is conserved." It seems to be conserved because we will it so!

But that is the way of science. Concepts are not retained because they are free of fault, perfectly infallible, or rigidly logical. *Theories are retained because they are useful* as we attempt to understand our observational knowledge and as we struggle to develop reliable expectations for the future.

Strange to say, built upon this sandy foundation, we find thermodynamics—the most certain basis for scientific prediction we have at our disposal! The reason for this confidence is that thermodynamics is really based upon empirical pillars that reach down through the sand of our theories to the bedrock of our observational experience. Thermodynamics embodies the Law of Conservation of Energy with confident certainty, not because of the elegance of our mathematical expressions for *vis viva,* but because many thousands of men have tried to construct a *perpetuum mobile* (a perpetual motion machine) and all have failed! Men were finally forced to conclude that no one would *ever* invent a machine that could generate work (energy) without consuming fuel. This massive accumulation of experience is our real basis for believing that energy must be conserved. This empirical foundation is as statistically incontrovertible as is the continued profitable existence of the gambling houses at Monte Carlo. One does not need a theory of roulette to conclude that the house always wins. And statistics brings us to another concept of thermodynamics, that of entropy.

About steam engines and efficiency

Nicolas Carnot was 28 years old and an army engineer when he published his classic work on the efficiency of steam engines. It was 1824 and he still thought of caloric as a substance. His interest was in the manner in which the steam in a cyclic heat engine repetitiously draws heat from a firebox, pushes back a
1820 piston to perform work, transfers heat to a cold reservoir, and draws back the piston to begin the cycle anew. He perceived that all the change in such a cyclic process occurs in the surroundings—the net effect is to transfer caloric (heat) from the firebox to the cold reservoir and to produce work. He considered the reversal of one heat engine driven by another, both operating between the same two heat reservoirs. He concluded that the maximum efficiency of any heat engine must be independent of the working fluid (steam, air, alcohol vapor, etc.) unless a perpetual-motion machine of a new kind were possible:

> ". . . if it were possible that the caloric should produce, by any process whatever, a larger quantity of motive power than that produced in our first series of operations, it would be possible . . .

to effect a return of caloric . . . from the refrigerator to the source [firebox]—and thus to re-establish things in their original state . . . there would thus result not only the perpetual motion, but an indefinite creation of motive power without consumption of caloric or of any other agent whatever. Such a creation is entirely contrary to the ideas now accepted . . . it is inadmissable.'' *

Some forty years later, Rudolph Clausius saw the deep significance of Carnot's work, but with the advantage that he viewed heat as the random energy of motion of the now-accepted atomic
1860 particles of matter. He continued Carnot's considerations of the heat effect q in a cyclic process operating at maximum efficiency. From these studies he concluded that a special significance could be ascribed to the quotient, the heat at maximum efficiency divided by the reservoir temperature, q/T. He was so confident of the significance of this quantity that he gave it a name, *entropy,* a name deliberately selected to sound like *energy* to emphasize its parallel importance. Then he proposed an addendum to the energy-conservation principle, an addendum that became known as the Second Law of Thermodynamics.

First Law of Thermodynamics:
The energy of the universe is constant.

Second Law of Thermodynamics:
The entropy of the universe tends toward a maximum.

We see that entropy is a sign post, indicating the direction that changes will take! What is this entropy? Ludwig Boltzmann told us in 1877.

Entropy and probability

Boltzmann was concerned with the kinetic theory of gases. He found that he could account for changes in gases and their states of equilibrium with a statistical model. In a paper with a catchy title, ''Über die Beziehung zwichen dem zweiten Hauptsatze der mechanischen Wärmetheorie und der Währscheinlichkeitsrechnung, respective den Sätzen über das Wärmegleichgewicht,'' he gave conceptual meaning to entropy.

''The belief is therefore expressed that we can calculate the
1875 state of equilibrium by investigating the probability of the different possible states of the system. The initial state will in most cases be a very improbable one and from it the system will progress toward more probable states, until it at last reaches the most probable state, that is, that of equilibrium. . . . If we apply this to the second law we can identify that quantity which we

* *Ibid.*, p. 244.

commonly designate as entropy with the probability of the actual state."*

Now the scene is set for Chemical Thermodynamics!

Thermodynamics in chemistry

The central problem in chemistry is the control and understanding of chemical change. Some key questions that must be answered are as follows:

- —When two substances are mixed, will they react?
- —If reaction occurs, will the reaction be accompanied by energy release?
- —If reaction begins, at what composition of reactants and products will reaction cease and equilibrium be established?
- —If reaction can occur, how rapidly will it proceed?

Chemical thermodynamics is concerned with the first three questions. It neither explains nor predicts the rate of a chemical reaction, the substance of the fourth question. It does, however, provide us with an understanding of the factors that determine whether a given mixture of substances has a spontaneous tendency to react (rapidly or slowly) to form other substances. It tells us the conditions that prevail when chemical changes no longer occur and chemical equilibrium has been reached. It does these things through considerations of the heat effects in reactions. Once again we find our beginnings in studies of caloric.

Hess' law and chemical reaction heats

As remarked in the first sentence of this prologue, man has ever been concerned with thermal effects. Consequently the chemistry of combustion has always been high in the chemists' list of interests. In fact, the industrial revolution was implemented with chemical fuels as the prime power source. No wonder that an early question brought to the laboratory was the measurement of the heat effects that accompany chemical change. Germaine Henri Hess gave this field its initial impetus by accurately measuring many reaction heats. His repertoire included a number of reactions that, taken in sequence, gave an overall chemical change that could also be studied directly. He drew from these cases a generalization which he published in 1840:†

"A combination having taken place, the quantity of heat evolved is always constant whether the combination is performed directly
1840 or whether it takes place indirectly and in different steps."

This law, once called "Hess' Law of Constant Heat Summation," but more meaningfully called the "Law of Additivity of Reaction

* *Ibid.*, p. 263.

† H. M. Leicester and H. S. Klickstein, *A Source Book in Chemistry*, p. 331, McGraw-Hill, New York, 1952.

Heats," is a chemical expression of the First Law of Thermodynamics. It merely asserts that the energy difference between reactants and products is the same whether the reaction proceeds directly or whether the products are reached through a series of reactions. Already, thermodynamics is in evidence.

Search for the cause of spontaneous change

Two workers who independently continued the accurate measurements of reaction heats, in particular, heats of combustion, were Marcellin Berthelot and Julius Thomsen. Berthelot introduced the term *exothermic* for a reaction that releases heat and *endothermic* for a reaction that absorbs heat. Both Thomsen and Berthelot, in their preoccupation with heat effects, proposed that the direction of a spontaneous reaction is determined by the heat evolved. Berthelot, a brilliant scientist, proposed the following in 1860:*

"*Principle of maximum work.* All chemical changes, occurring
1860 without the intervention of outside energy tend toward the production of bodies, or of a system of bodies, which liberate more heat."

The erroneous nature of this principle was patently obvious even when proposed. Otherwise Berthelot would not have needed to define the term "endothermic." Many processes that occurred spontaneously with the absorption (not liberation) of heat were known at the time, even the household experience that salt dissolving in water cools the solution! Nevertheless, there seemed to be a correlation between the enthusiasm of many chemical reactions and their heat liberation; so this inadequate principle received a certain acceptance.

Along came Gibbs

In a classic work published around 1875, J. Willard Gibbs laid the solid foundation for all of chemical thermodynamics. With the advent of the particulate view of matter, Gibbs recognized the applicability of the powerful methods of statistical mechanics; he appreciated the significance of Boltzmann's words about probability and change. He proceeded to develop these into a
1875 formalism that was glorious in its abstraction and formidable in its intricacy. In terms of energy-related quantities symbolized by Greek symbols ϵ, χ, ζ, and ψ, the most diverse physico-chemical phenomena could be considered; the factors that determine spontaneous change were implicit. The symbols chosen by Gibbs were most appropriate. His work remained "Greek" to practically all chemists for another quarter of a century.

**Ibid*, p. 431.

Nernst and another law of thermodynamics

But the search for the origins of spontaneous change went on, practically unaided by the esoteric work of Gibbs: most of it had to be discovered again. One of the important contributors was W. Nernst, who also understood the pronouncements of Boltzmann about equilibrium and the most probable state. Hence he was prepared, in 1906, to recognize the significance of certain conclusions offered by T. W. Richards concerning the temperature dependence of the voltages and heats generated in electro-
1900 chemical cells. These conclusions led Nernst to propose that the entropy of every substance becomes zero at the absolute zero of temperature. This became known as the Third Law of Thermodynamics, and it implies that every substance at any other temperature has a positive entropy, an *absolute* entropy relative to this absolute zero reference point.

Lewis and the opening of chemical thermodynamics

It remained for Gilbert N. Lewis, working at the University of California in Berkeley, to bring chemical thermodynamics into the everyday thinking of every chemist. More than any other single scientist, he and his many students showed how to apply thermodynamics to chemical processes. In particular, Lewis fostered
1920 the use of the free energy concept, both as a means for predicting the direction of spontaneous change and as a quantitative indicator of the equilibrium state. He designated free energy *F*, and identified it with Gibb's quantity ζ (zeta).* His profound influence on chemistry was guaranteed when, in 1923, the textbook *Thermodynamics,* by G. N. Lewis and M. Randall, was published, bringing to the scientific world the insight accumulated at Berkeley.

There could be no more fitting proof of the impact of chemical thermodynamics than the fact that it has become the cornerstone of the introductory course in chemistry.

The aim of this monograph

The qualitative ideas of chemical thermodynamics provide the framework within which chemistry now operates. Even without the powerful mathematical thermodynamic relationships, the myriad of known chemical facts are made coherent and meaningful with the aid of the conceptual ideas. So we shall not attempt to present all, or even a large part, of the rigor of thermodynamics. Instead, we shall attempt to convey the *meaning* of thermo-

*In singular disrespect, many chemists use neither Gibb's symbol ζ nor Lewis's symbol *F*, but rather designate free energy by the symbol *G*.

Note added in proof: A qualified critic labelled this footnote "an irreverent oversimplification of many years of negotiation at the international level." To this indictment, the authors plead guilty and, in un-Berkeley-like submissiveness, we shall use the symbol *G* for free energy throughout this monograph.

dynamic principles. In particular, we shall attempt to help the reader feel comfortable with the quantity, entropy. With Clausius, we believe the entropy concept to be every bit as important in our thinking as the energy concept. Hopefully, this presentation will elucidate the way in which entropy points the way for spontaneous
1969 change. We hope it will do this, both for those who would be in a foreign land if the language of calculus were used, and also for those who might miss intuitive significances if the mathematical complexities were presented in all their glory.

To accomplish this goal we have made one significant sacrifice. In many places we have explored meanings in terms of microscopic visualizations (what are the molecules doing?). This detracts from thermodynamics in one of its most important strengths. Because thermodynamics is founded in the empirical observations of macroscopic behavior, its verity and power do not *require* a microscopic model. That is why Carnot was able to provide the basis of the Second Law of Thermodynamics, while still viewing caloric as substantive. We believe this sacrifice will be more than repaid if a new generation of chemists is started on its way thinking intuitively in conformity with the secure guidelines given us by thermodynamics.

one
a microscopic view of equilibrium

One of the greatest strengths of thermodynamics is that it is entirely based on empirical observations on the macroscopic (bulk) level. Its validity and its predictions do not require any microscopic (molecular) view of the process under study. A sure way to put across this message is to present the thermodynamic reasoning and to apply it without ever referring to molecules and their individual behaviors. Most books on thermodynamics pursue this laudatory path.

We have consciously decided to sacrifice this course in favor of a presentation that parallels the thermodynamic logic with a chemist's view of the microscopic happenings. Our primary aim is to give correct intuitive meaning to the principles we consider, and it can best be done by appealing to our highly developed picture of the life and times of molecules.

So this first chapter is devoted to a thermodynamic concept of the greatest importance—the nature of the equilibrium state—but we will consider it initially in a mechanistic, non-thermodynamic way. This view of "what is going on" at equilibrium will provide a valuable intuitive setting for the subsequent chapters which pursue the thermodynamic description of equilibrium.

1-1 The equilibrium state

In mechanics, an equilibrium state is one of balance between opposing forces. It is recognized by an appearance of quiescence—nothing is changing as time passes. Let us see how much of this description we can transfer to a chemical system.

(a) SOME CHEMICAL SYSTEMS: WHICH ARE AT EQUILIBRIUM?

Figure 1-1 shows three systems for which a number of observations indicate a constancy of properties. The first is a bunsen burner flame. In a quiet room, the flame appears quite immobile; a thermocouple placed at a particular place inside the flame or near the flame indicates a constant temperature. Yet everyone recognizes that a bunsen-burner flame is not an equilibrium situation. The reason is that gas and air are continuously being added to the flame at the base of the burner, and hot carbon dioxide and water are ejected at the top of the

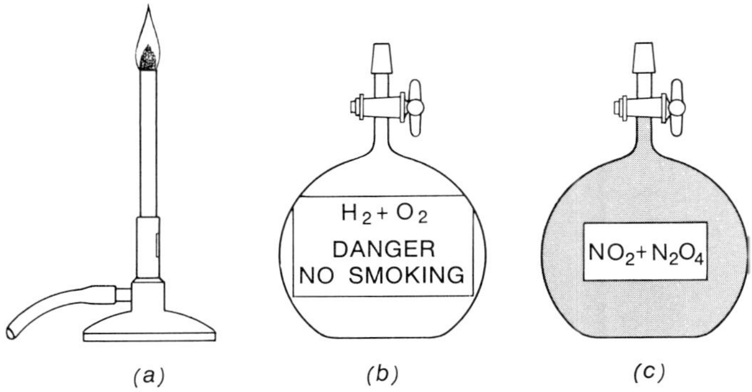

Figure 1-1 Systems with constant properties: Which are at equilibrium?

flame. *Equilibrium can exist only in a closed system:* a system to which neither energy nor material substances are being continuously supplied or removed. The constant addition of reactants, and removal of both products and heat, means the flame is not a closed system, so it cannot be at equilibrium. A system that achieves constancy of properties in this way is called a *steady state.*

The second system in Figure 1-1 is a bottle containing a mixture of hydrogen and oxygen. Again, it is common knowledge that such a mixture is explosive and only the tiniest spark is needed to initiate violent proof that it is not at equilibrium, chemistry-wise. The reason the mixture will just sit there, looking at you innocently, is that the explosive reaction $2H_2 + O_2 \rightarrow 2H_2O$ is extremely slow under normal conditions of temperature and pressure. Once the reaction gets started, however, the hydrogen and oxygen molecules know what to do! Nevertheless, this gives us a second caveat to place on the use of the constancy of properties as a means of recognizing an equilibrium situation. *The rate of any possible change that might occur must be noticeably rapid. Then* the constancy of properties implies equilibrium.

The third system in Figure 1-1 is a bottle containing a mixture of NO_2 and N_2O_4. The bottle is colored because one of the constituents, NO_2, is colored. The system *is* at equilibrium with respect to the reaction

$$2NO_2(g) \rightleftarrows N_2O_4 \qquad (1\text{-}1)$$

How do we know? By three simple tests. *First,* we verify that it is a closed system. It is: the bottle is tightly stoppered and no heat is being transferred. *Secondly,* we measure properties that depend on the equilibrium and see that they do not vary over time. For example, as reaction (1-1) proceeds, the pressure drops (two molecules of NO_2 produce only one of N_2O_4) and the orange color of NO_2 is reduced in intensity. However, neither the pressure nor the color is changing. *Thirdly,* we make a small and momentary change in some condition that is known to affect the reaction in question. For example, we could first lower the temperature a few degrees, and then restore it to its initial value. Immediately we would perceive a slight decrease in the orange color of the NO_2, and then a restoration to the original color. The reaction proceeds a bit to the right, producing more N_2O_4 and consuming a bit of NO_2, and then, as the temperature is raised again, the original color is restored. *This proves the reaction in question proceeds rapidly and reversibly under our conditions of study.* Therefore we have an equilibrium system, and we have shown how to identify it.

A system is at equilibrium when:

- *neither energy nor material substance is being added or removed (it is a closed system);*
- *the measurable bulk properties of the system are not changing with time;*
- *possible changes proceed at a measurable pace, as revealed by the response to a small disturbance which causes observable properties to change and then to return to their initial values when the disturbance is removed.*

(b) A CLOSER LOOK AT EQUILIBRIUM

Despite the generally placid appearance of chemical systems at equilibrium, scientists are convinced that things are not as quiet on the atomic level. In fact, it is found experimentally that atoms are constantly in motion. In the gas phase, molecules dash about, bouncing off the walls of the container. Molecules in liquids and solids are not this active but they are, nevertheless, constantly jiggling and vibrating. Keeping these characteristics in mind, let's consider a very simple system—a closed vessel containing a little liquid water—and watch how it attains equilibrium.

When the system is at equilibrium, there will be some water vapor present along with the liquid water; the exact amount will depend only on the temperature. The gas-phase molecules will be dashing around, hitting both the sides of the flask *and* the surface of the liquid. Will they have a chance of sticking in the liquid, or will the original group of molecules that evaporate stay in the gas phase for as long as the conditions remain the same? This can be tested quite easily using the apparatus in Figure 1-2, which consists of two identical flasks joined by a stopcock. Each contains liquid water, but the right-hand flask contains some radioactive water, HTO. (The symbol T is used to represent tritium, the hydrogen isotope of mass number three, ${}^{3}_{1}H$.) Both flasks have the same temperature and water vapor pressure. When the stopcock is opened, the vapors mix, and soon equal numbers of radioactive molecules are present in the gas in each container. What about the liquid in the left-hand flask? If, after a time, some is removed and tested with a Geiger counter, it will be found that this flask contains some radioactive water. A plot,

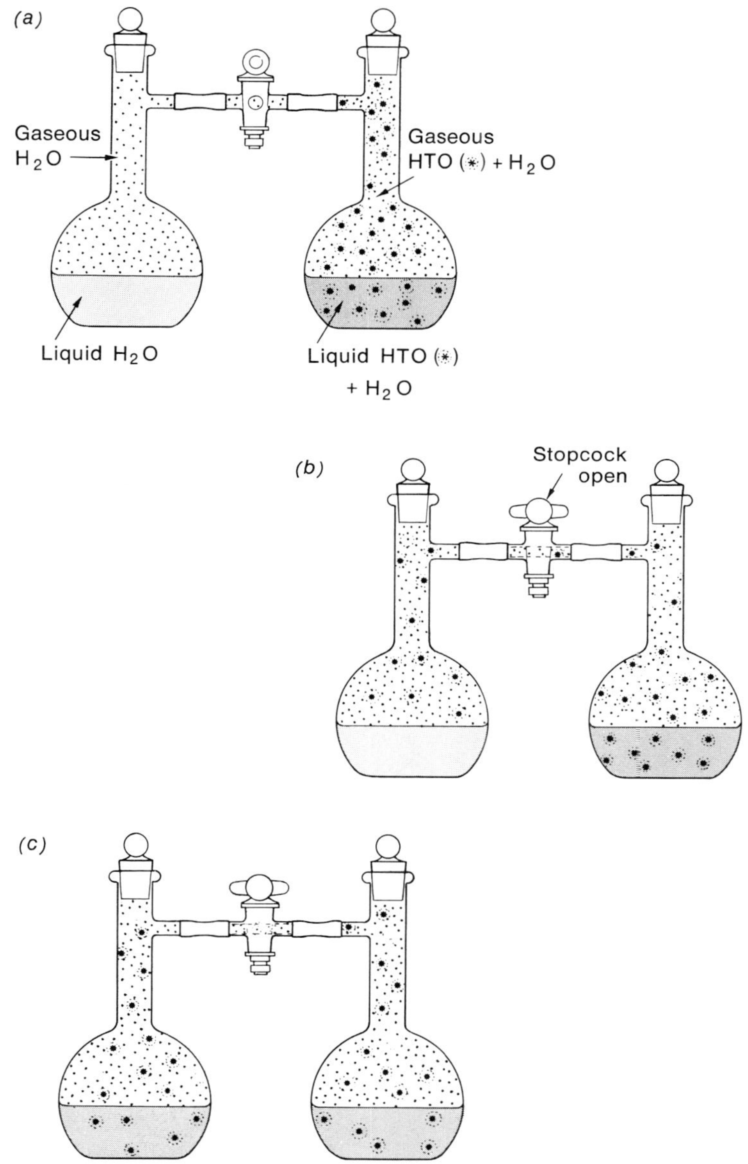

Figure 1-2 The dynamic nature of equilibrium. (a) *Both flasks contain H_2O, and the right-hand flask contains some radioactive HTO ($T \equiv {}^3_1H$).* (b) *When the stopcock is opened, the vapors mix; though the pressure remains constant, gaseous HTO passes into the left-hand flask.* (c) *After a time, if liquid is removed from the left-hand flask, it too will exhibit radioactivity; again, while pressure remains constant, HTO condenses.*

against time, of the radioactivity in the liquid phases in each flask would look like Figure 1-3. Thus, we must conclude that molecules are constantly entering and leaving the surface of the liquid. Equilibrium is not a static process, but a dynamic one.

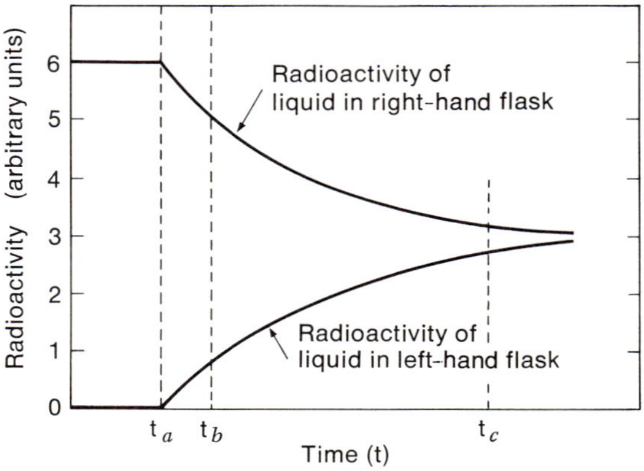

Figure 1-3 The vapor–liquid equilibrium is dynamic. At t_a, when the stopcock is opened, all HTO radioactivity is in the right-hand flask. A short time later, at t_b, HTO is observable in the left-hand flask, though the vapor pressure remains constant. A long time later, at t_c, the HTO is almost uniformly distributed between the liquids in the two flasks, though the vapor pressure has not changed all the while.

For our liquid $\rightleftarrows$ vapor equilibrium, we must have the condition

$$\text{rate of evaporation} = \text{rate of condensation}$$

or

$$\begin{bmatrix}\text{number of molecules}\\ \text{leaving surface of}\\ \text{liquid per second}\end{bmatrix} = \begin{bmatrix}\text{number of molecules}\\ \text{entering surface of}\\ \text{liquid per second}\end{bmatrix}$$

A simple calculation shows that, at room temperature, there are about 10^{22} molecules evaporating per second from each square centimeter of the water surface. For the vapor pressure to remain constant, this same huge number of gaseous H_2O molecules must be condensing each second on that same square centimeter of surface. Thus, we see that the constancy of observable properties is actually achieved by a whirlwind of activity at the molecular level. **Equilibrium occurs,** not when all activity ends, but **when the rates of opposing processes become equal.**

(c) THE EQUILIBRIUM CONSTANT IN A PARTICULAR REACTION

A chemist mixes hydrogen and iodine in the gas phase at 700°K. Reaction begins, as shown by a decreasing intensity of the color of the purple iodine gas; hydrogen iodide is produced.

$$H_2(g) + I_2(g) = 2HI(g) \qquad (1\text{-}2)$$

After a time, the iodine color stops decreasing—equilibrium has been reached. The remaining color shows that some reactants remain unconsumed. Now, one of the most important questions a chemist can ask about the system is, how far did it go? Were most of the reactants consumed, so that now the bulb contains mainly HI? Or was only a small percentage of the iodine converted?

One possible way to answer this question would be to examine the ratio

$$\frac{\text{concentration of products at equilibrium}}{\text{concentration of reactants at equilibrium}} \qquad (1\text{-}3)$$

This number would be large if reaction (1-2) proceeds almost to completion and small if it proceeds barely at all. In fact, we could look at the ratio for various mixtures to see how it depends on the relative concentrations. For this purpose, we might examine the quotient

$$\frac{(\text{conc. HI})}{(\text{conc. } H_2)(\text{conc. } I_2)} \qquad (1\text{-}4)$$

Table 1-1 shows, in the fourth column, the magnitude of ratio (1-4) for five different mixtures. We see that the number is high (hydrogen iodide is favored) and it depends upon the mixture composition.

Table 1-1 Concentration Relations for H_2–I_2 Mixtures at Equilibrium, $T = 698.6°K$

Experimental Results			Calculated Ratios	
$[H_2]$ mole/liter	$[I_2]$ mole/liter	$[HI]$ mole/liter	$\frac{[HI]}{[H_2][I_2]}$	$\frac{[HI]^2}{[H_2][I_2]}$
1.8313×10^{-3}	3.1292×10^{-3}	17.671×10^{-3}	$3.08 \times 10^{+3}$	54.5
2.9070	1.7069	16.482	3.32	54.6
4.5647	0.7378	13.544	4.02	54.4
0.4789	0.4789	3.531	15.4	54.4
1.1409	1.1409	8.410	6.47	54.4

Many decades ago, chemists learned empirically that for a reaction such as (1-2), the most informative way to represent the equilibrium situation is to modify the ratio (1-3) by squaring the concentration of a constituent that appears in the balanced equation with a coefficient 2, cubing the concentration of a constituent with a coefficient 3, and so on. Thus, our quotient (1-4) would be better written as

$$\frac{(\text{conc. HI})^2}{(\text{conc. H}_2)(\text{conc. I}_2)} \tag{1-5}$$

With the same equilibrium concentrations, Table 1-1 lists the values of quotient (1-5) in the last column. Now the ratios are always the same! Apparently we now have a constant number that characterizes equilibrium at the temperature of study. This number is called the equilibrium constant K:

$$K = \frac{(\text{conc. HI})^2}{(\text{conc. H}_2)(\text{conc. I}_2)}$$

or, as concentrations are usually symbolized by brackets, we write

$$K = \frac{[\text{HI}]^2}{[\text{H}_2][\text{I}_2]} \tag{1-6}$$

(d) THE EQUILIBRIUM CONSTANT FOR ANY REACTION

In the last section we have merely reminisced, for a particular case, on the empirical discovery of the Equilibrium Law. For a generalized reaction between reactants $A,B,\ldots$ to give products $C,D,\ldots$, the balanced equation would be

$$wA + xB + \cdots \rightleftarrows yC + zD + \cdots \tag{1-7}$$

and the generalized concentration relationship would be

$$\boxed{K = \frac{[\mathbf{C}]^y[\mathbf{D}]^z \cdots}{[\mathbf{A}]^w[\mathbf{B}]^x \cdots} \qquad \textbf{The Equilibrium Law}} \tag{1-8}$$

Experimentally we find that the concentration quotient (1-8) (including the integer exponents $w,x,y,$ and z) turns out to be a constant at a given temperature. The constant is symbolized K and is called the equilibrium constant. Once we measure this number (experimentally), we can calculate the end result, that is, the equilibrium situation, for any set of starting conditions. Obviously K is an important and useful quantity; one we must understand.

We shall see later that the magnitude of the equilibrium constant K can be understood in thermodynamic terms. In this chapter we will examine the magnitude of K in mechanistic terms.

1-2 Equilibrium and reaction rate

The Equilibrium Law, (1-8), can be derived for a given reaction merely by equating the rate of the reaction in the forward direction to its rate in the reverse direction. Though the derivation is simple, scientists with purer souls object to it. Both the argument and the objection are educational, so we will review them both, using the real reaction (1-2) as an example.

(a) THE RATE OF FORMATION OF PRODUCTS

We'll begin with a mixture of H_2 and I_2, and consider first the formation of HI. If we try to describe on a microscopic scale the events leading to formation of HI, it is natural to picture a hydrogen molecule colliding with an iodine molecule and, in an exchange of atoms, forming two HI molecules (see Fig. 1-4). Some of the factors that fix the rate at which this process will occur are obvious. Clearly, the rate depends upon collisions between H_2 and I_2 molecules if our picture is correct. Therefore, the rate to the right is proportional to the concentration of each constituent that must participate in the collision, and, in turn, the concentrations are determined by the pressures p_{H_2} and p_{I_2}.

$$H_2 + I_2 \xrightarrow{(\text{rate})_R} \text{products} \tag{1-9}$$

$$(\text{rate})_R \text{ is proportional to } (p_{H_2}) \times (p_{I_2}) \tag{1-10}$$

Since HI neither helps nor hinders the collisional process (1-9), its partial pressure does not affect $(\text{rate})_R$, so it does not appear in (1-10). The proportionality constant k_R, in (1-10), includes lots of information we chemists would like to decipher. It tells us what fraction of the H_2–I_2 collisions are effective, how important the collisional geometry is at the time of encounter, and how the rotational and the vibrational excitations affect the atomic rearrangement. Furthermore, it contains the temperature dependence of the process pictured in Figure 1-4. In our ignorance of all these details, we merely write

$$(\text{rate})_R = k_R(p_{H_2})(p_{I_2}) \tag{1-11}$$

(b) THE RATE OF FORMATION OF REACTANTS

As reaction (1-2) proceeds, hydrogen iodide accumulates. Then, occasionally, two hydrogen iodide molecules will collide, and a fraction of these collisions will disrupt the HI molecules to form H_2 and I_2. This process to the left occurs at a rate limited by the frequency of HI–HI collisions, as pictured in Figure 1-5:

$$\text{reactants} \xleftarrow{(\text{rate})_L} \text{HI} + \text{HI} \tag{1-12}$$

Because two HI molecules are involved in the collision, raising the pressure of HI is doubly effective. There are not only more HI molecules to collide, but also more to collide with. Hence, the

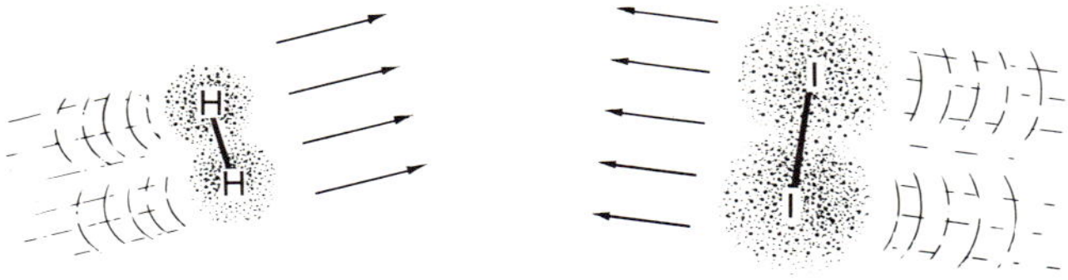

(a) Watch out, here it comes!

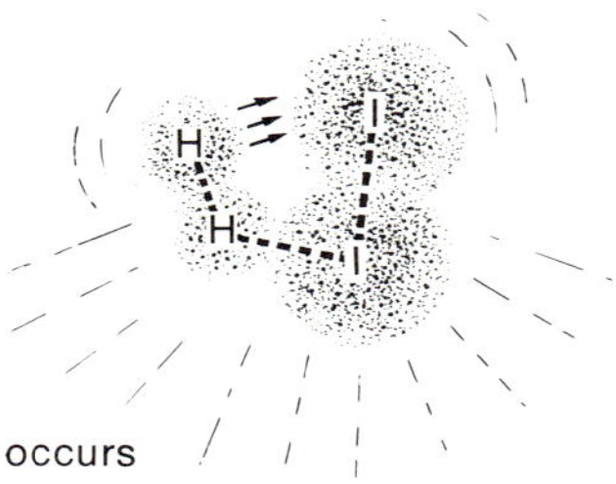

(b) Blam!
The collision occurs

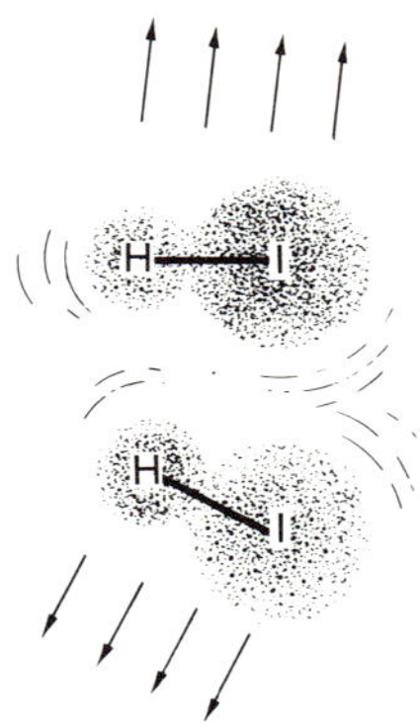

(c) Look what happened!

Figure 1-4 A reactive collision between a hydrogen molecule and an iodine molecule to form hydrogen iodide.

rate to the left is proportional to the square of the HI pressure:

$$(\text{rate})_L \text{ is proportional to } (p_{HI})^2 \qquad (1\text{-}13)$$

Once again, there is a fascinatingly interesting proportionality constant k_L that concerns the detailed dynamics of collisions such as those pictured in Figure 1-5. Knowing little about these dynamics, we merely write

$$(\text{rate})_L = k_L(p_{HI})^2 \qquad (1\text{-}14)$$

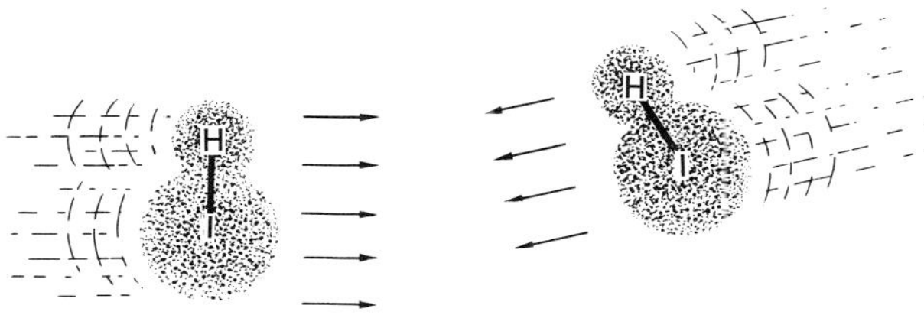

(a) Brace yourself!

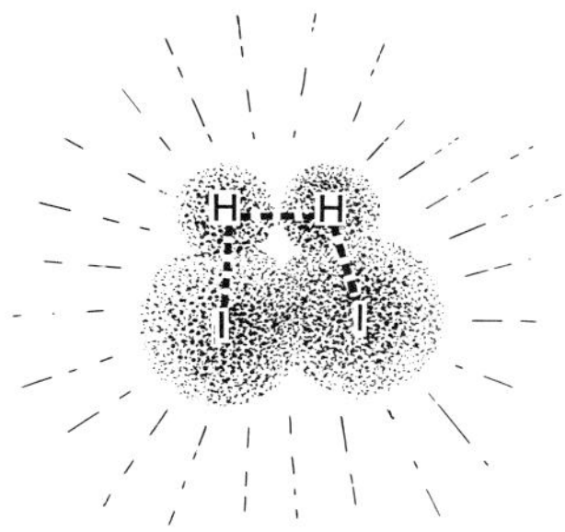

(b) Ouch!

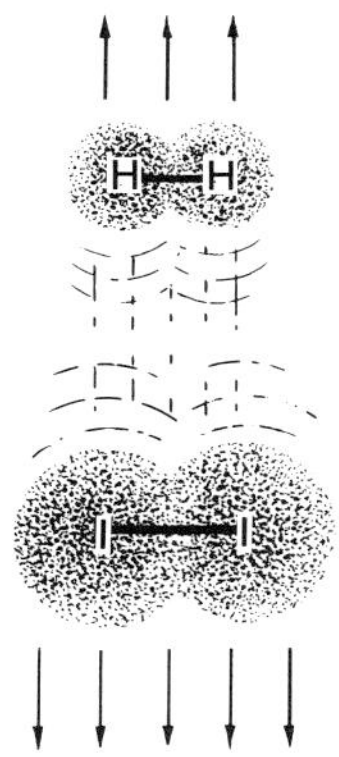

(c) Now look what happened!

Figure 1-5 Hydrogen iodide molecules can collide too, to give back hydrogen and iodine.

(c) EQUILIBRIUM: A BALANCE BETWEEN OPPOSING REACTIONS

When H_2 and I_2 are first mixed, the rate of formation of HI, reaction (1-9), is rapid, and since no HI is present initially, the rate of reforming H_2 and I_2 by the reverse reaction (1-12) is zero. As the reaction proceeds, however, both H_2 and I_2 are used up, so $(\text{rate})_R$ slows down. Meanwhile, the HI is accumulating, so $(\text{rate})_L$ is increasing. As this process continues, $(\text{rate})_R$ dropping and $(\text{rate})_L$

rising, the two reaction rates will approach each other and finally become equal. At this point the rate of formation of HI is exactly counterbalanced by an equal rate of loss of HI through the reverse reaction. Then, despite the continued reaction to right and left, a dynamic balance exists, and concentrations no longer change. We call this equilibrium. It is characterized as follows:

$$\text{At equilibrium} \qquad (\text{rate})_{\rm R} = (\text{rate})_{\rm L} \tag{1-15}$$

Substituting (1-11) and (1-14) into (1-15),

$$k_{\rm R}(p_{\rm H_2})(p_{\rm I_2}) = k_{\rm L}(p_{\rm HI})^2 \tag{1-16}$$

Rearranging,

$$\frac{k_{\rm R}}{k_{\rm L}} = \frac{(p_{\rm HI})^2}{(p_{\rm H_2})(p_{\rm I_2})} \tag{1-17}$$

However, the ratio of two constants must also be a constant, so $k_{\rm R}/k_{\rm L} = K$. Comparing the pressure relationships in (1-17) to the concentration relationships in (1-6), we see that equating forward and reverse reaction rates leads to the Equilibrium Law*

$$K = \frac{(p_{\rm HI})^2}{(p_{\rm H_2})(p_{\rm I_2})} \tag{1-18}$$

(d) WHY SOME LIKE IT AND SOME DON'T

This reaction rate derivation of the Equilibrium Law has great merit. It conveys a dynamic picture of the equilibrium state that every chemist believes exists. Although an equilibrium mixture seems static and quiescent from a macroscopic view, the chemist looks into it with his intuitive submicroscopic vision and sees lots of molecular reactions going on; reactants meeting to give products, and, at the same time, products meeting to give reactants. He also sees that the static condition represents a state of balance between these opposing processes. This dynamic view of equilibrium, and the approach to equilibrium, is an extremely valuable one.

So what's the objection? If all chemists—including the purest of them—believe equilibrium is dynamic, and that it involves equality of opposing rates, why do some of them object to the reaction rate derivation of the Equilibrium Law? The answer is that the rate derivation requires a microscopic view of *how* the reaction proceeds, whereas the thermodynamic argument does

*Notice that in (1-18) the proportionality factor between pressure and concentration occurs squared in the numerator and squared in the denominator, hence it cancels. In such a case, the numerical magnitude of K is independent of the units selected; all quantities can be in pressure units, e.g., atmospheres, or all can be in concentrations, e.g., mole/liter or mole/cm^3. This is not true in general, for usually the number of molecules consumed is not equal to the number of molecules produced.

not. As mentioned earlier, one of the greatest strengths of thermodynamic deductions is that they are independent of a detailed understanding of chemical processes at the molecular level. We can debate—profitably, mind you—whether or not the H_2–I_2 reaction mechanism is as simple as shown in Figures 1-4 and 1-5. In fact, this worry merits some detailed consideration.

(e) BACK TO THE RATE DERIVATION—WHAT IF OUR MECHANISM IS WRONG?

Fortunately, it can be shown that the rate derivation of the Equilibrium Law is independent of the actual mechanism. The H_2–I_2 reaction is a fine case in point, since the mechanism has been challenged only recently. John H. Sullivan recently proposed that the reaction, which proceeds rapidly only at elevated temperatures, actually involves an equilibrium, thermal dissociation of iodine atoms and a series of subsequent reactions as follows:*

$$I_2(g) \rightleftarrows 2I(g) \tag{1-19}$$

$$I(g) + H_2(g) \rightleftarrows HI(g) + H(g) \tag{1-20}$$

$$H(g) + I_2(g) \rightleftarrows HI(g) + I(g) \tag{1-21}$$

Sullivan further postulates that reaction (1-20) is much slower than reaction (1-21). This means that whenever (1-20) occurs (consuming an iodine atom), (1-21) immediately follows (replacing the iodine atom).

These deductions, which pertain to the rate of the reaction when no hydrogen iodide is present, can be applied to the equilibrium situation. Equilibrium is, of course, characterized by constant composition. Hence, we can conclude that the rate at which H_2 is being consumed must be equal to the rate at which it is being produced. All the arguments used in the derivation (1-15) to (1-18) can be applied to (1-20):

$$I(g) + H_2(g) \underset{k_{L,20}}{\overset{k_{R,20}}{\rightleftarrows}} HI(g) + H(g) \tag{1-22}$$

where

$$\text{rate of loss of } H_2 = k_{R,20}(p_I)(p_{H_2})$$

and

$$\text{rate of production of } H_2 = k_{L,20}(p_{HI})(p_H)$$

At equilibrium,

$$k_{R,20}(p_I)(p_{H_2}) = k_{L,20}(p_{HI})(p_H)$$

*J. H. Sullivan, *Journal of Chemical Physics,* **Vol. 46,** p. 73, 1967.

hence,

$$\frac{k_{R,20}}{k_{L,20}} = K_{20} = \frac{(p_{HI})(p_H)}{(p_{H_2})(p_I)} \qquad \textit{(1-23)}$$

However, if reaction (1-20) does not cause either net production or net consumption of H_2, then it will cause no net production or consumption of H atoms. This means that the constancy of composition at equilibrium *requires* that reaction (1-21), the only other source of H atoms, must *also* neither consume nor produce H atoms. Once again

$$H(g) + I_2(g) \underset{k_{L,21}}{\overset{k_{R,21}}{\rightleftarrows}} HI(g) + I(g) \qquad \textit{(1-24)}$$

where

$$\text{rate of loss of H} = k_{R,21}(p_H)(p_{I_2})$$

and

$$\text{rate of production of H} = k_{L,21}(p_{HI})(p_I)$$

At equilibrium,

$$k_{R,21}(p_H)(p_{I_2}) = k_{L,21}(p_{HI})(p_I)$$

hence,

$$\frac{k_{R,21}}{k_{L,21}} = K_{21} = \frac{(p_I)(p_{HI})}{(p_H)(p_{I_2})} \qquad \textit{(1-25)}$$

We now have two equilibrium relationships among pressures: (1-23) and (1-25). Neither is easy to verify experimentally because each involves the partial pressures of hydrogen atoms and of iodine atoms. These two constituents, because of their reactivity, are present at very minute concentration, too low to measure. We can, however, combine the two expressions in a way that eliminates the undetectable intermediates. Multiplying (1-23) by (1-25) gives

$$K_{20} \times K_{21} = \frac{(p_{HI})(p_H)}{(p_{H_2})(p_I)} \times \frac{(p_I)(p_{HI})}{(p_H)(p_{I_2})}$$

By cancelling, and recognizing that the product of two constants is itself a constant, $K_{20} \times K_{21} = K$, and hence,

$$\boxed{K = \frac{(p_{HI})^2}{(p_{H_2})(p_{I_2})}} \qquad \textit{(1-18)}$$

This is exactly the result that has been observed experimentally and that was derived by assuming another kinetic mechanism.

This is not an accident. *Any possible reaction mechanism will lead to the same equilibrium relationship, the Equilibrium Law*! In fact, a thermodynamic argument can be framed that requires this to be so (the "Principle of Microscopic Reversibility").* So there is, after all, no objection to deriving the Equilibrium Law by equating the opposing rates, as long as it is realized that the correct outcome does not imply that the mechanism assumed must be the most important one occurring. We are on a one-way street. Since the same Equilibrium Law is obtained for any mechanism, we can derive this law by equating the rates of the opposing reactions for the most likely mechanism. Having obtained the correct Equilibrium Law relationship, we cannot, however, conclude that this gives any assurance that the mechanism is the one actually doing the work. All mechanisms lead to the same law.

1-3 Equilibrium: A state of dynamic balance

Since this mechanistic view of equilibrium is in agreement with the empirical Equilibrium Law, we are inclined to regard it as a valid model. In fact, it is generally accepted today that, at equilibrium in a chemical reaction, both reactants and products are present. The bulk properties of the system do not change because reactants are forming products at exactly the same rate at which the products are reforming the reactants. The position of equilibrium will favor products if the forward rate is intrinsically rapid compared with the reverse rate. Even then, there is some concentration level of the reactants which is sufficiently small that, finally, the forward and reverse reaction rates become equal. Equilibrium will be established.

An immediate consequence of this picture is that the release of energy cannot alone account for reaction tendency. The proposal of Berthelot, quoted in the Prologue, that reactions always proceed so as to liberate energy, is foredoomed. The reaction between H_2 and I_2 releases energy; yet the reaction between two HI molecules *does* take place and, as Table 1-1 shows, does so sufficiently to cause the equilibrium mixture to contain substantial amounts of all constituents, H_2, I_2, and HI.

Nevertheless, the energy effects in chemical reactions are extremely important to us. In fact, we shall see that the *thermodynamic understanding of spontaneous change, and of equilibrium, requires that we consider the energy changes involved.* So, recognizing that the position of equilibrium is not determined solely by energy changes, and that even endothermic reactions will proceed spontaneously, let us turn to the investigations of reaction heats.

*The argument is again based on the empirical impossibility of the perpetual motion machine that could be made by coupling two mechanisms with different Equilibrium Laws for the same reaction; one proceeding in the forward direction and the other proceeding in the reverse direction.

Fuels burning in a rocket engine produce great quantities of heat and rapidly expanding gases that push huge satellites into outer space. Food consumed by animals is transformed by the body into energy which can be used by the muscles in a multitude of daily tasks. A chemistry student in the lab may notice as he dissolves salt in water that the solution becomes markedly cooler. An electrical engineer designing an electric car needs a battery that will provide electrical energy as cheaply as gasoline provides mechanical energy in an internal combustion engine.

It is clear that energy changes are a crucial part of our day-to-day existence, and it is not surprising that the study of these energy effects is an important part of chemistry. Before beginning a study of the thermodynamics of energy changes, we will focus our attention on events on the molecular scale* to increase our understanding of the origins of these energy effects.

two
energy and chemical change

2-1 Molecular energies

(a) KINETIC ENERGY

Energy of motion is called kinetic energy. In physics we learn that the kinetic energy of a body of mass m traveling with a velocity v is given by

$$KE = \tfrac{1}{2}mv^2 \tag{2-1}$$

Kinetic energy resulting from the overall motion of a body is called energy of translation ($KE_{\text{trans.}}$). An orbiting satellite, a speeding automobile, a gas molecule, an atomic electron—all have translational kinetic energy.

What about motions other than translational? If we think carefully about our speeding automobile, we recognize that the car is alive with moving parts: pistons are going up and down, wheels are going around, valves are opening and closing. Each of these makes its own contribution to the *total* kinetic energy of the moving automobile. We might write

$$(KE)^{\text{auto}}_{\text{total}} = KE_{\text{trans.}} + KE_{\text{pistons}} + KE_{\text{wheels}} + KE_{\text{valves}} + \cdots \tag{2-2}$$

*Throughout this book we will use the word "microscopic" to describe events involving individual atoms or molecules, and "macroscopic" to describe events involving large numbers of atoms or molecules. Thus, a reaction involving a mole of gas would be called a macroscopic system.

Some of these contributions might be small compared with the overall translational motion, but they still must be considered in a complete treatment. Some of them continue when the car is parked, if the engine is running.

Like automobiles, atoms and molecules have motions that are more complex than simple translations. A gaseous helium atom, for example, has two negatively charged electrons moving in the field of the positive nucleus. Each electron has kinetic energy of its own, so, for the total kinetic energy of a helium atom, we must write

$$(\mathrm{KE})_{\mathrm{He}} = \mathrm{KE}_{\mathrm{trans.}} + \mathrm{KE}_{\mathrm{electron\ 1}} + \mathrm{KE}_{\mathrm{electron\ 2}} \qquad (2\text{-}3)$$

The kinetic energy of the electrons is maintained even when the helium atom is "parked," that is, when it is at rest, so that $\mathrm{KE}_{\mathrm{trans.}} = 0$.

Molecules are even more complicated. In addition to the translational kinetic energies of the molecule and of the individual electrons, there are two further possibilities. In the gas phase, we must consider vibrations of the atoms relative to each other, and rotation of the molecule as a whole. The three contributions from the atomic nuclei to the overall kinetic energy are illustrated in Figure 2-1 for the XeF_2 molecule. For the total kinetic energy of any molecule, we can write

$$\mathrm{KE}_{\mathrm{molecular}} = \mathrm{KE}_{\mathrm{trans.}} + \mathrm{KE}_{\mathrm{electrons}} + \mathrm{KE}_{\mathrm{vibration}} + \mathrm{KE}_{\mathrm{rotation}} \qquad (2\text{-}4)$$

Of course, all these motions go on simultaneously, and even a simple molecule like XeF_2 would appear to be a busy little object were we able to see it.*

In solids and liquids, the treatment of kinetic energy is not as simple as in the gaseous case discussed above. Tight packing and strong forces between molecules constrain the molecules to back-and-forth movement about fixed positions. The molecular vibrations are about the same, but the translations and rotations found in the gas phase are generally obstructed in condensed phases.

(b) POTENTIAL ENERGY

Picture a car parked on the side of a steep hill (Fig. 2-2). Should the brake suddenly fail, experience tells us that the car will spontaneously start to roll downhill, gathering momentum as it goes. With a radar trap at the bottom to measure its velocity, we could calculate its kinetic energy of translation. Where did this energy come from? In order to satisfy the belief that energy is not created or destroyed in ordinary processes, but merely transformed, a

*The student may question the confidence with which we can attribute properties to an invisible object. In fact, the measurements of molecular spectroscopy show that the picture of molecular motions given above is very accurate indeed.

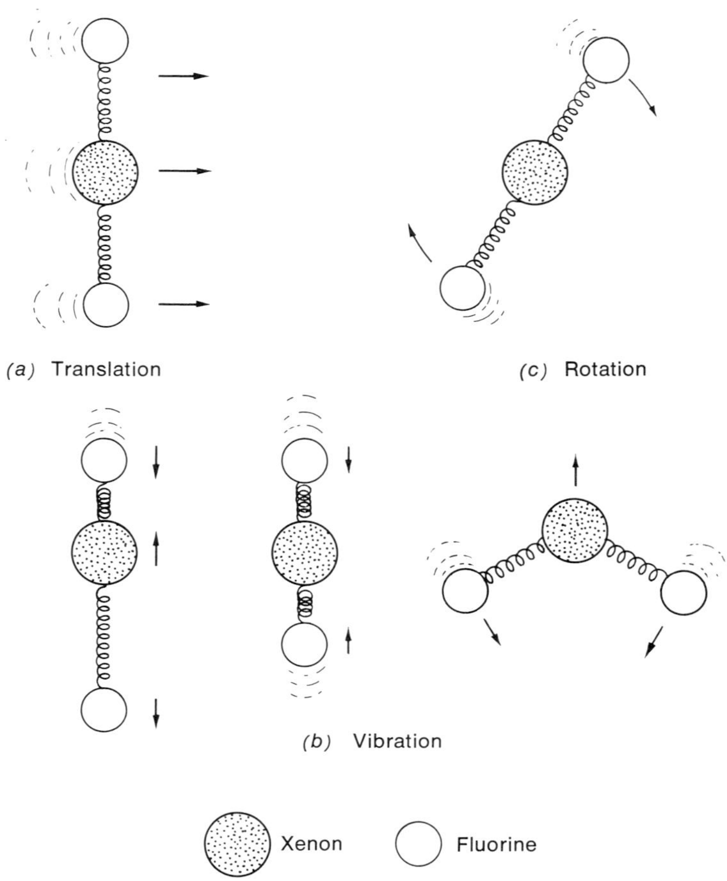

Figure 2-1 Nuclear motions of $XeF_2(g)$

new kind of energy was invented to deal with this question. In its original position, the car was "potentially" able to roll downhill and to gather kinetic energy as it went. Hence, we attribute to the car at the hilltop some stored energy that can be released when needed. The stored energy associated with its position on the hill is called *potential* energy. Detailed experiments would show that the final kinetic energy achieved by the car would be proportional to its original position atop the hill, or, in general,

$$PE = mgh \tag{2-5}$$

where h is the height above the bottom, and g is the gravitational constant.

Atoms and molecules have potential energy also. Atoms in molecules are held together by forces related to the electrical attraction between positive nuclei and negative electrons. In a bonding situation, these attractive forces are exactly balanced

Figure 2-2 (a) *The car has potential energy only;* (b) *the car rolls spontaneously downhill;* (c) *all potential energy has been converted to kinetic energy.*

by repulsive forces between like charges on pairs of electrons and pairs of nuclei. Thus, potential energy of position in molecules results from electrical interactions. Energy released or absorbed during chemical processes results from the making and breaking of chemical bonds and the resulting changes in this potential energy.

We can summarize the sources of potential energy in molecules as

$$\mathrm{PE}_{\text{molecule}} = \mathrm{PE}_{\text{electron-nucleus}} + \mathrm{PE}_{\text{electron-electron}} + \mathrm{PE}_{\text{nucleus-nucleus}} \tag{2-6}$$

$\mathrm{PE}_{\text{electron-nucleus}}$ is associated with the attractive electron–nuclear interaction

$\mathrm{PE}_{\text{electron-electron}}$ is associated with the repulsive electron–electron interaction

$\mathrm{PE}_{\text{nucleus-nucleus}}$ is associated with the repulsive nuclear–nuclear interaction

We have not yet mentioned the enormous amounts of energy stored in atomic nuclei as a result of attractive forces between the nuclear particles, neutrons and protons. (The nature of these forces is not fully understood even today.) The energies involved in ordinary chemical reactions are never sufficient to interfere with the nuclear bonding, so changes in nuclear potential energy do not normally take place. Systems in which changes in nuclear bonding do occur can release vast amounts of energy—such processes account for the energy coming from the sun, the work supplied by a nuclear power plant, and the fearsome destructiveness of a hydrogen bomb.

(c) TOTAL ENERGY

The total energy of a molecule is given by a summation of the potential and kinetic energy terms we have been discussing:

$$E_{\text{total}} = \mathrm{KE} + \mathrm{PE}_{\text{molecule}} + \mathrm{PE}_{\text{nuclei}} \tag{2-7}$$

It should be clear that the measurement of E_{total} is a formidable task for one molecule, and a prodigious one in the case of a mole of molecules (one mole contains 6×10^{23} molecules). The branch of science that concerns itself with the calculation of absolute energies is called statistical mechanics or statistical thermodynamics.

The chemist, however, often does not need to know the *absolute* energy, E_{total}, of what he has in his beakers. Usually it suffices if he knows only the *change* in energy that takes place during the chemical processes that occur. He wants to know if the reaction absorbs heat so that the beaker has to be heated while the reaction is in progress, or, conversely, if heat is released, so that the reaction could serve as an energy source for some useful purpose. This more limited need simplifies things. All we have to do is find a way to measure the energy released or absorbed during the chemical process. We can escape from worrying about the exact manner in which the energy is distributed among the molecules of reactant and the molecules of product. The quantity of interest is

$$\text{energy change} = \Delta E$$

where (2-8)

$$\Delta E = [E_{\text{total}}\ (\text{products})] - [E_{\text{total}}\ (\text{reactants})] \quad (2\text{-}9)$$
$$= [(\text{KE} + \text{PE})_{\text{products}}] - [(\text{KE} + \text{PE})_{\text{reactants}}] \quad (2\text{-}10)$$

Since the nuclear energy does not change during the chemical reactions of interest to us, the term $\text{PE}_{\text{nuclei}}$ of equation (2-7) is the same for both reactants and products, and does not appear in (2-10).

(d) WHY DO WE CARE?

If chemical thermodynamics, the subject of this monograph, does not need to know the allotment of energy among the molecular claimants, why have we bothered to discuss molecular vibrations, rotations, electron kinetic energy, and the like? Why do chemists care? The answer is simple. Mere measurement of the quantities of energy exchanged during chemical processes does not bring with it any understanding of what is actually going on at the molecular level. Chemists, striving to explain and predict chemical reactions, find they need to know the nature of the changes involved, as well as their magnitudes. Yet, *a microscopic picture of chemical events is not required in chemical thermodynamics.*

2-2 Energy conservation

Energy is conserved. No single theory of physics is more widely accepted or more generally useful, yet the statement refers to an abstract concept about a quantity never measured directly. We measure velocity and mass to calculate energy of motion. We measure an altitude (from which valley?) to determine energy of position. We measure moles of a substance to infer its chemical energy. We measure the change in the density of mercury to infer transfer of heat. Frequently, the main evidence for the existence of a quantity or type of energy is that energy is apparently not conserved unless some unseen energy is assumed. A classical example was the postulate of an undetected nuclear particle, the neutrino, in the beta decay of the nucleus, which saved the energy-conservation law. A more far-reaching instance was Einstein's deduction that $E = mc^2$, which requires that substance (mass) must also be recognized as energy.

To understand and work with this abstract and many-faceted idea, we will examine some simple processes. We will see how energy is observed and how it is recognized in different forms during these processes. First, however, we will need a simple vocabulary.

(a) THE LANGUAGE OF THERMODYNAMICS

We almost always find it convenient to consider and study the properties of a limited part of our environment—one-half mole of ammonia gas in a 2-liter bulb, 200 milliliters of 1.5 molar potassium permanganate solution in a beaker, an eight-cubic-foot refrigerator and its contents. We call such a limited part of the things around us a *system.* If these limits are clearly stated and carefully controlled, another individual can study a like part of our environment, that is, another one-half mole of NH_3 in another 2-liter bulb, and so on. Then he can duplicate our measurements and observations to check our results, and extend them to add to our knowledge. This limiting of our concern to a clearly stated part of the environment lends simplicity to our study and precision to our conclusions. If a system is not able to interact with any other system, we say it is *isolated.* All other systems with which our particular system might exchange energy are called the *surroundings.* (For example, the surroundings of our system might be a constant-temperature bath in which our 2-liter bulb of NH_3 is immersed; or a hot plate on which the beaker of permanganate is placed, plus the atmosphere around it; or, if we can't be more specific, the surroundings of a system can always be said to be the rest of the universe.)

(b) THE EXCHANGE OF ENERGY

Consider the simple mechanical system illustrated in Figure 2-3. When the billiard ball is struck by the cue, energy is exchanged. In the formal language of physics, work is done on the ball in propelling it (hopefully) on its intended path. Energy has been transferred to the ball. At the same time, the player's hand is slightly warmed by friction as the cue rubs his fingers. Some of the energy put into the cue by the player is returned to him, through friction, as heat. Energy is being exchanged, some as heat and the rest as work. These are the two forms of energy transfer that will concern us in our studies of chemical thermodynamics.

(c) HEAT AS ENERGY

We are quite used to the idea that energy dissipated through friction is lost as heat. However, only about 150 years ago, the

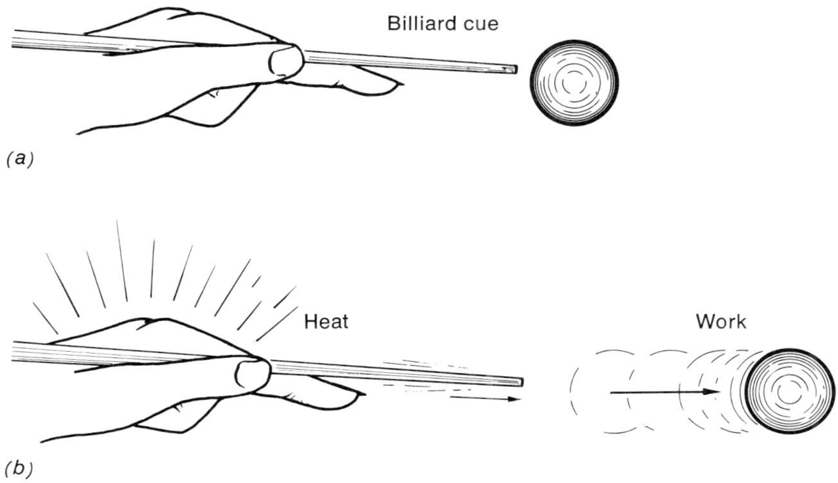

Figure 2-3 Exchange of energy. (a) *Energy is stored in the muscles of player;* (b) *energy exchanged as heat (friction on hand) and work (on ball).*

view was that heat was a material substance called caloric which was transferred from a hotter to a colder body. The science of thermodynamics, which is the study of heat, only began after Count Rumford observed in 1798 that an apparently inexhaustible amount of heat was produced when a cannon was bored. It seemed strange to him that a fixed amount of iron could contain apparently limitless quantities of caloric. These doubts led him to perform a controlled experiment in which the heat produced by a blunt borer was transmitted to a vessel of cold water. He observed:

> ". . . it was not long before the water which surrounded the cylinder began to be sensibly warm. . . . At two hours and twenty minutes it was at 200; and at 2 hours 30 minutes it ACTUALLY BOILED! . . . It is hardly necessary to add, that any thing which any *insulated* body, or system of bodies can continue to furnish *without limitation* cannot possibly be a *material substance:* and it appears to me to be extremely difficult, if not quite impossible, to form any distinct idea of any thing, capable of being excited and communicated in the manner the Heat was excited and communicated in these Experiments except it be MOTION." *

*W. F. Magie, *A Source Book in Physics,* pp. 158–161. Harvard University Press Cambridge, 1963.

Now, over 150 years later, we find it difficult to appreciate the astonishment with which Rumford perceived the results of his observations. What is astonishing, however, is his intuitive conclusion that heat is *motion.* It is easy for us with our sophisticated equipment and detailed knowledge of atomic and molecular activity to understand that an increase in temperature is just the bulk manifestation of an increase in kinetic energy of the particles of a body. Rumford did not even dream of atoms and molecules!

(d) THE FIRST LAW OF THERMODYNAMICS

The First Law of Thermodynamics is merely the statement that energy is conserved during any change. Energy is neither created nor destroyed, but simply changed in form. Rumford's work shows that it is necessary to include heat within the boundaries of the energy-conservation law.

In the case of a chemical process, this tells us that whenever a certain reaction occurs, such as the combustion of propane gas,

$$C_3H_8(g) + 5O_2(g) \rightarrow 3CO_2(g) + 4H_2O(g) \qquad (2\text{-}11)$$

the difference between the potential energy held by the reactants (1 mole propane and 5 moles oxygen) and that held by the products (3 moles carbon dioxide and 4 moles water vapor) will result in energy being either released or absorbed. The amount of energy depends only on the chemical nature of the reactants and products. Some of the energy may be released as heat and some as work (perhaps work done in pushing back the atmosphere), but, for a fixed amount of propane, the sum of the heat and the work will always be the same. This can be expressed quantitatively as follows:

$$\Delta E = E_{\text{final}} - E_{\text{initial}} \qquad (2\text{-}12)$$

where ΔE is the change in energy in going from the initial to the final state.

$$\boxed{\Delta E = q - w} \qquad (2\text{-}13)$$

w = work *done by* system
q = heat *absorbed by* system

The need for the minus sign is clear when we realize that work done by the system implies a loss of energy, while heat absorbed implies energy again. Thus defined, the two quantities must be summed with opposite sign. The reason for this particular convention in signs is an historical one. Steam engineers, who pioneered the systematic study of thermodynamics, expressed the law in terms of the quantities they were accustomed to measuring. Naturally, they were interested in the heat that had to be put into a system (say, to heat a boiler) and in the work that would be delivered in return.

(e) MUST ENERGY BE CONSERVED?

The First Law is a statement encompassing the results of a vast accumulation of experimental evidence. We cannot say that energy *must* be conserved—however, we do know that, in the total experience of science to date, energy always *has been* conserved. To understand how our experience teaches us to believe in the First Law, we need only consider the effects of a violation of this law. Engineers have long hunted diligently for an engine that would perform more useful work than the energy supplied to make it run. Such an engine would, of course, violate the First Law, for it would "create" energy. The hunt was spurred on by the knowledge that such an engine would be more rewarding than a slot machine that clanged out two quarters every time one was dropped in. Our conviction of belief in the First Law is founded in the knowledge that such an engine has never been found—although not through want of trying. After all the past effort, we are convinced that it will be the same in the future. Energy having always been conserved before provides a basis for expectation in experiments yet to be done. This is the route that scientific advance must take. We diligently accumulate a set of consistent facts, we build our theories to fit these facts, and then we base our expectations on these theories.

(f) CHEMICAL WORK

Energy can be transferred from one system to another either by means of heat flow or by means of work. It is important to bear in mind that the energy E is a characteristic property of a system (the summation of the potential and kinetic energy terms). Work and heat flow are *not* properties of a system but they are the *means* by which energy is transferred when a change occurs.

There are many varieties of work, but in chemistry we are mainly concerned with electrical work and work done by expanding gases. Electrical work can be produced by an electrochemical cell. Expansion work results from a change in volume of the systems involved and is usually called pressure–volume work, or PV work. Expanding gases in the cylinder of an automobile engine do work on the piston; this work is harnessed to turn the wheels. Expanding gases from an open reaction vessel do work in pushing back the atmosphere; this work is lost to us. All kinds of work, whether useful or not, must be considered if our energy bookkeeping is to come out right.

Mechanical work is done when a force moves through a distance, and it is expressed numerically by

$$\begin{aligned} \text{work} &= (\text{impressed force}) \times (\text{distance moved}) \\ &= f \cdot (\Delta r) \end{aligned} \tag{2-14}$$

We can investigate pressure–volume work by considering what happens when gas contained in a cylinder is allowed to expand, pushing back a piston against some external force f_{ex}. A simplified diagram of an apparatus to perform this experiment is shown in Figure 2-4.

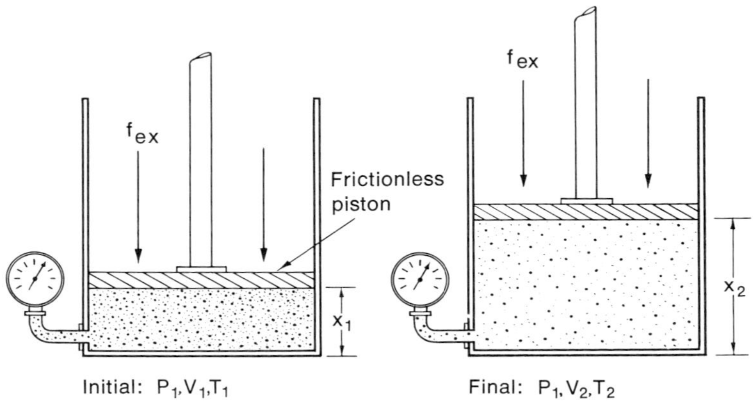

Figure 2-4 Pressure–volume work against constant external force (such as atmospheric pressure). The gas pressure as registered on the gauge remains constant throughout the process.

At the beginning of the expansion, the force due to the internal pressure of the gas just equals the external force. Now, if we gently heat the cylinder, the internal pressure will increase and the piston will move until the force exerted on it by the gas again balances the external force. In Figure 2-4(*b*), the piston has moved from its original position x_1 to a new position x_2. The work done, as the piston pushed against the constant external force, is

$$\text{work} = f_{ex} \cdot (\Delta x) \tag{2-15}$$

We may express this equation in terms of more easily measured quantities (always a primary aim in science) if we divide *and* multiply the right-hand side of (2-15) by *A*, the cross-sectional area of the piston:

$$\begin{aligned} \text{work} &= f_{ex} \cdot (\Delta x) \cdot \frac{A}{A} \\ &= \frac{f_{ex}}{A} \cdot [A \cdot (\Delta x)] \end{aligned} \tag{2-16}$$

If we perform our experiment so that f_{ex} always exactly balances the internal pressure of the gas, the first term in (2-16), f_{ex}/A (the force per unit area), is just the internal pressure of the gas, P. The second term $[A \cdot (\Delta x)]$ is the change in volume of the cylinder during the expansion, ΔV. Thus, (2-16) becomes

$$\begin{aligned} \text{work} &= \frac{f_{ex}}{A} \cdot [A \cdot (\Delta x)] \\ &= P \cdot \Delta V \\ &= P(V_2 - V_1) \end{aligned} \tag{2-16a}$$

The First Law of Thermodynamics can now be rewritten to include $P\Delta V$ (the explicit expression for PV work) and w' (all other forms of work).

$$\boxed{\Delta E = q - P\Delta V - w'} \tag{2-17}$$

or, if $w' = 0$

$$\boxed{\Delta E = q - P\Delta V} \tag{2-17a}$$

This is the First Law in its most useful form to a chemist.

A graphical representation of PV work is shown in Figure 2-5(*a*).

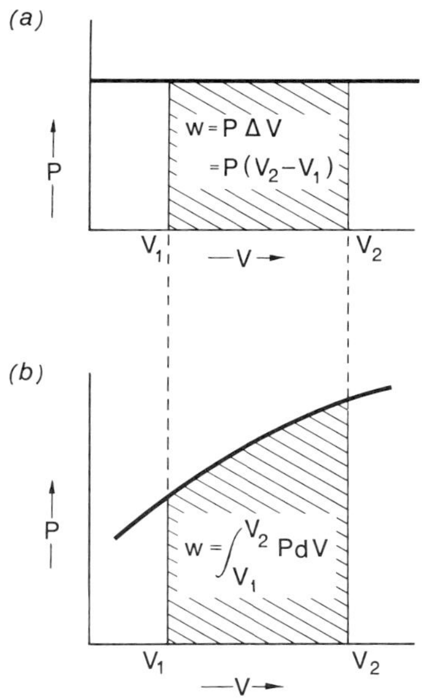

Figure 2-5 Pressure–volume work is given graphically by the area under P–V *plot.* (a) *Constant pressure,* $w = P\Delta V$; (b) *pressure not constant, area given by the integral* $\int_{V_1}^{V_2} P dV$ *(see Appendix A).*

When the pressure is kept constant, as in the above example, the *PV* work done is given numerically by the shaded area on the graph. Part (*b*) of this figure shows an example of an expansion during which the pressure was not constant. The work done (in going from V_1 to V_2) is still given by the magnitude of the shaded area. However, this value is not given by the simple expression (2-17), but must be worked out using the methods of integral calculus. A detailed knowledge of calculus is not necessary at this stage. It is sufficient to recognize that it helps us calculate the shaded area in Figure 2-5(*a*) and (*b*). The process of calculating this area, the area under the curve, is called *integration,* and it is symbolized by the integral sign $\int$. The upper bound of the shaded area is fixed by the curve *P,* and the limits of the area are the initial and final volumes, V_1 and V_2. The shaded area, which is numerically equal to the work done, is then expressed as

$$w = \int_{V_1}^{V_2} PdV \qquad (2\text{-}18)$$

The term "*dV*" is merely another expression for ΔV in which the volume difference has become extremely small. Expression (2-18) is read aloud as "the work done equals the integral of *PdV* between the limits $V = V_1$ and $V = V_2$." This process is more completely discussed in Appendix A.

(g) ENERGY, A FUNCTION OF STATE

Consider the situation shown in Figure 2-6. In part (*a*), one ball is sitting at the top of a hill, with a potential energy proportional to the height of the hill. If it is released, it rolls to the bottom, bashes into the other ball, and comes to a halt. All of its potential energy has been dissipated; some as heat, through frictional

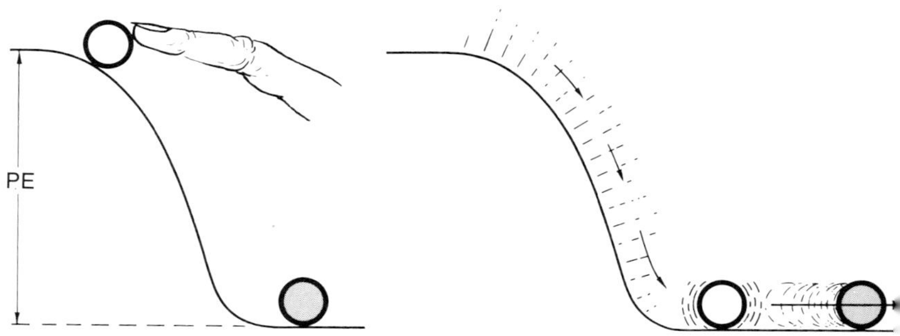

(*a*) (*b*)

Figure 2-6 Energy is a state function. Potential energy, PE, *of the unshaded ball in* (a) *is completely transferred to hill (heat) and to the other ball (work) in* (b).

interaction with the hill, and some as work performed on the second ball. Our experience in these situations, as summed up in the First Law, tells us that the total amount of energy dissipated as work and heat will exactly equal the initial potential energy of the ball, measured relative to the bottom of the hill. If we were to polish the hill, thus reducing friction, more work would be done and less heat produced. *The total amount of energy exchanged would be the same.* In fact, any process which results in the ball moving from the top to the bottom of the hill will dissipate the same amount of energy, *no matter how this energy is distributed between heat and work.*

The potential energy of a molecule depends on the relative positions of the electrons and nuclei of the various atoms that make up the molecule. When two molecules react to form a third, the atoms are rearranged, their electrons are placed differently relative to the nuclei, and the potential energy of the products differs (usually) from that of the reactants. Thus, the energy change ΔE depends only on the nature of the products and the reactants. As in the above example of the ball rolling downhill, any energy released may be dissipated as work or heat in varying proportions, but ΔE is always the same for a given chemical change.

Quantities which depend only on the state of the system (such as the energy E) are called *state functions.* Other state functions are pressure, temperature and volume. The importance of this state function concept is clear. If we once measure the energy change ΔE for some process, we know that *any other* process effecting the same change on our system will have the same energy change. This is true even though the relative amounts of heat and work exchanged with the surroundings might be vastly different. Only one measurement of ΔE, then, need ever be made for a particular initial and final state.

It is simple to demonstrate that PV work is not a state function. Consider the PV plots in Figure 2-7. Suppose we have 0.01 mole of gas at an initial pressure $P_1 = 200$ torr and volume $V_1 = 20$ ml and expand it to a final pressure $P_2 = 800$ torr and a final volume $V_2 = 80$ ml. This change can be brought about as in Figure 2-7(*a*), first increasing the pressure to its final value, $P_2 = 800$ torr, at constant volume V_1, by heating the gas. No work is done in this step, because $\Delta V = 0$. Then at constant pressure $P_2 = 800$ torr, the gas is expanded to the final volume V_2 by heating the gas (Fig. 2-7(*c*)). The work done is given by the shaded area and is equal to

$$\begin{aligned} w_1 &= P_2\Delta V \\ &= P_2(V_2 - V_1) \\ &= (800)\cdot(80 - 20) \\ &= 48{,}000 \text{ cc torr} \end{aligned} \tag{2-19}$$

In Figure 2-7 (*b*) we approach the same final state by a different

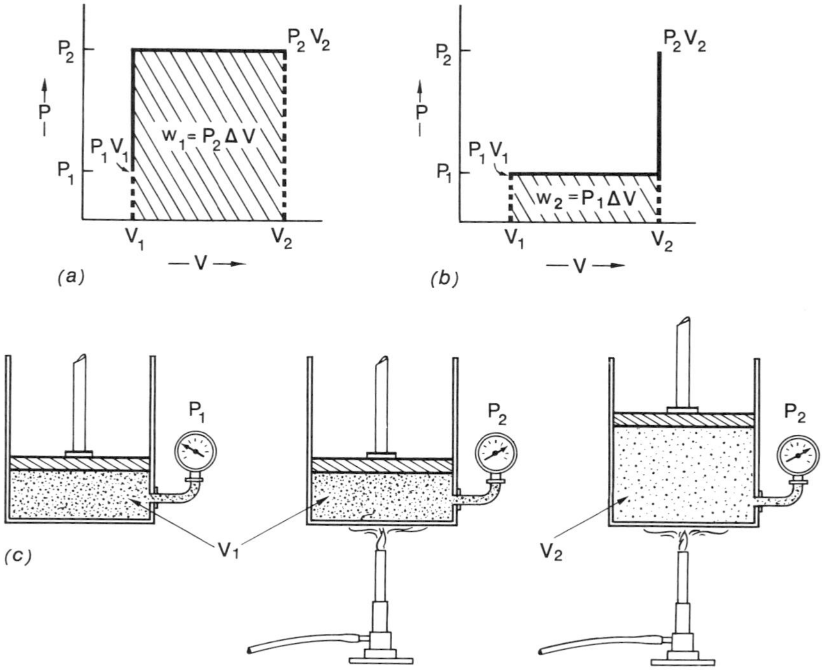

Figure 2-7 Pressure–volume work is not a state function. Gas in a cylinder is taken from an initial state P_1V_1 to a final state P_2V_2 by two different routes (a) *and* (b)*. The work done in the two cases is different, as shown by the shaded areas. Thus work done depends on the path. Part* (c) *illustrates the process described graphically in* (a)*. The gas in a volume V_1 and registering a pressure P_1 is heated. If the piston is held fixed, the pressure rises to P_2. At this point the piston is allowed to move, so that the pressure remains constant as the volume increases to V_2.*

route. The first step is a warming of the gas so it expands to the final volume V_2, at a constant pressure $P_1 = 200$ torr. After this step, the gas is warmed to increase the pressure at this constant volume until the final state P_2V_2 is reached. No work is done in the second step because the volume is constant. Hence the work done is given by the PV product in the first step, shown shaded on the graph:

$$
\begin{aligned}
w_2 &= P_1 \Delta V \\
&= (200) \cdot (80 - 20) \\
&= 12{,}000 \text{ cc torr}
\end{aligned}
\qquad (2\text{-}20)
$$

In these two examples the overall change was the same, yet the work done was different. In order to satisfy the First Law, it is necessary that q, the heat absorbed, also be different in these two cases. Only ΔE, the state function, is the same for both processes.

2-3 Thermochemistry

The chemical engineer designing a chemical plant worries about the heat needed to keep a reaction going, or the cooling necessary to absorb heat evolved by a reaction. From a practical point of view, then, the part of the energy change dissipated as heat is an important part. In the next chapter we shall see how the heat of reaction also plays a crucial role in determining the position of equilibrium, hence, the extent of reaction. Generally, the value of q depends on how the reaction is carried out. However, q itself becomes a state function when a reaction is carried out either at constant volume or at constant pressure, two conditions that are easily maintained in practice. In this section we will investigate the nature of q under these special conditions.

(a) DEFINITIONS AND UNITS

To make measurements of heat changes, we need some consistent unit. Quantity of heat transferred is usually measured in terms of the *calorie*. One calorie is defined as the amount of heat needed to raise the temperature of one gram of water from 14.5°C to 15.5°C.* In chemistry, the energies involved when a single molecule reacts are immeasurably small, so it is necessary to deal with a convenient number of atoms, or molecules; usually a mole (1 mole contains 6.023×10^{23} atoms or molecules). The heat evolved or absorbed during a chemical reaction will be expressed in calories per mole of reactant, or, if the numbers are large, kilocalories per mole of reactant (1000 cal = 1 kcal). A reaction that evolves heat is said to be *exothermic,* while one that absorbs heat is called *endothermic.*

(b) CONSTANT VOLUME PROCESSES

When a reaction is carried out in a closed vessel, the volume cannot change and the work done is zero; that is,

$$\Delta V = 0 \tag{2-21}$$

and hence

$$w = P\Delta V = 0 \tag{2-22}$$

The First Law expression (2-13) simplifies to

$$\begin{aligned} \Delta E &= q - w \\ &= q - 0 \\ &= q_V \end{aligned}$$

*For precision thermal measurements, a calorie is now defined with reference to electrical quantities: 1 calorie = 4.1840 joules.

The subscript V is used to indicate a constant volume process. In this case, the energy change ΔE, a function of state, is given by the heat released or absorbed during the process. It is important to realize that q is not generally a state function. It is equal to ΔE only when the process is carried out in one particular way—at constant volume. In Section 2-4 (p. 57), we will show how fixed-volume devices, known as "bomb calorimeters," are routinely used in the laboratory to measure q_V.

(c) CONSTANT PRESSURE PROCESSES

Much of chemistry is performed in beakers and other vessels that are open to the atmosphere. During any such process the volume of the system can change, but the pressure imposed by the atmosphere remains constant. Because of their great practical importance, much information on constant pressure processes is available in tabulated form.

Over one hundred years ago, J. Thomsen investigated the thermochemistry of the neutralization by sodium hydroxide of sodium bicarbonate and of carbon dioxide solutions. His experimental results were as follows:

$$\text{I} \qquad CO_2(aq) + 2NaOH(aq) \rightleftarrows Na_2CO_3(aq) + H_2O(\ell) \qquad q_P^{\text{I}} = -20.2 \text{ kcal/mole}$$

$$\text{II} \qquad NaHCO_3(aq) + NaOH(aq) \rightleftarrows Na_2CO_3(aq) + H_2O(\ell) \qquad q_P^{\text{II}} = -9.2 \text{ kcal/mole}$$

What about the direct conversion of CO_2 to $NaHCO_3$ as shown in equation III?

$$\text{III} \qquad CO_2(aq) + NaOH(aq) \rightleftarrows NaHCO_3(aq) \qquad q_P^{\text{III}} = ?$$

Thomsen estimated q_P^{III} in the following way. Consider the flow diagram Figure 2-8. The first step on the lower route to $NaHCO_3$ is reaction I. The second step is the reverse of reaction II. If reaction II is exothermic by 9.2 kcal/mole, the reverse reaction must *absorb* an identical amount of heat, or $-q_P^{\text{II}}$. The overall heat of reaction at constant pressure for the two-step process is thus q_P^{III}, since the final result is the conversion of CO_2 to $NaHCO_3$. This is shown as a one-step process by the top arrow. Therefore

$$\begin{aligned} q_P^{\text{III}} &= q_P^{\text{I}} - q_P^{\text{II}} \\ &= -20.2 - (-9.2) \\ &= -11.0 \text{ kcal/mole} \end{aligned}$$

With this argument Thomsen predicted that reaction III should be exothermic by −11.0 kcal/mole. Shortly thereafter, M. Berthelot verified this result experimentally by measuring q_P^{III} directly.

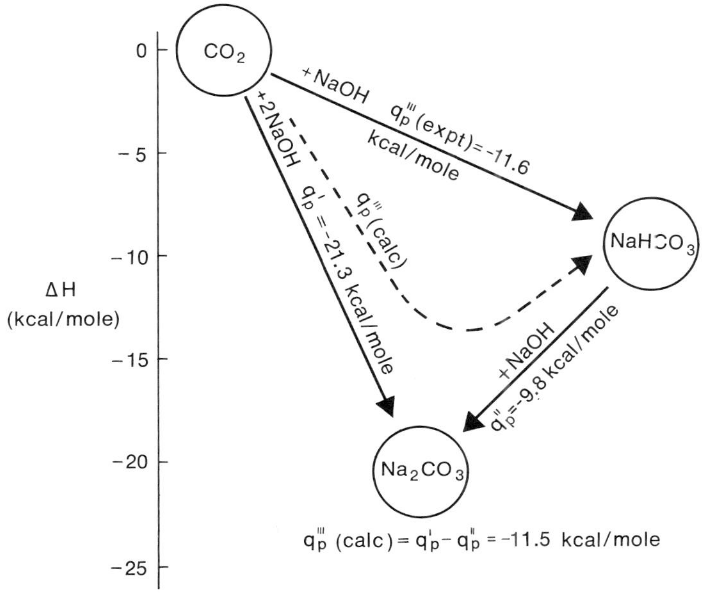

Figure 2-8 The value of q_p^{III} was first obtained from the experimental results for the two-step process in which Na_2CO_3 is an intermediate. This value agrees very well with the best experimental value for q_p^{III} presently available.

As a result of this experiment, we can conclude that the heat of reaction at constant pressure for the conversion of CO_2(aq) into sodium bicarbonate is the same for the two different paths investigated. A vast amount of experimental data on countless reactions permit the generalization that q_P is always independent of path; that is, *q_P is a state function.* This result allows an enormous reduction in the amount of experimental work that need be performed. As long as some stepwise route between products and reactions can be devised, in which q_P is already known for each step, q_P of the overall process can be evaluated without further experiment. We shall devote most of the rest of this section to investigating the usefulness of this result.

(d) ENTHALPY

Let us now solve the First Law expression, 2-13, for q_P.

$$\Delta E = q - w \qquad (2\text{-}13)$$

so, at constant pressure,

$$\begin{aligned} q_P &= \Delta E + w \\ &= \Delta E + P\Delta V \end{aligned} \qquad (2\text{-}23)$$

or, when written out in full,

$$q_P = (E_2 - E_1) + (PV_2 - PV_1) \quad (2\text{-}24)$$

A simple rearrangement gives

$$q_P = (E_2 + PV_2) - (E_1 + PV_1) \quad (2\text{-}25)$$

So we see that the heat associated with a process carried out at constant pressure can be given by a difference between two terms, each having the form

$$E + PV \quad (2\text{-}26)$$

We have already noted the experimental observation that q_P is a state function. Hence we must be able to express it as the difference between some state property in the final and initial states. Expression (2-25) shows that q_P is given by the difference in the value of the function $(E + PV)$ evaluated at the final and initial states. Because of the importance of constant pressure processes, this function (2-26) is given the special symbol H, and is called the *heat content* or *enthalpy* of the system. For any constant pressure process, the heat evolved or absorbed is given by ΔH, the enthalpy change during the process.

$$\Delta H = q_P = H_2 - H_1$$

where

$$\boxed{\mathbf{H = E + PV}} \quad (2\text{-}27)$$

We see that H, observed to be a state function, can be expressed in terms of E, P and V, each of which is a state function. In general any function made up entirely of state functions is itself a function of state.

(e) HEAT OF FORMATION

Because the enthalpy H is a function of the state of the system only, we can calculate ΔH for any process, as long as we can find a route for which ΔH is known for each step. It is this result, based solidly on the sort of experimental data described in Section 2-3(c), that makes thermodynamics so useful. A relatively small number of experimental measurements permits the calculation of ΔH for vast numbers of different reactions. For convenience, the data are tabulated according to reaction type: heat of combustion, heat of vaporization, heat of fusion, etc. A particularly useful process to consider is the *formation* of the compound of interest from its constituent elements. The enthalpy change for this process is called the heat of formation and it is symbolized ΔH_f. For example, the heat of formation of water vapor is the heat evolved when the following reaction occurs:

$$H_2(g) + \tfrac{1}{2}O_2(g) \rightleftarrows H_2O(g)$$

$$\text{heat of formation} = \Delta H_f(H_2O) \tag{2-28}$$

So that this number can be understood by every scientist, some standard set of experimental conditions must be specified. Heats of formation refer to a reaction carried out at 25°C (room temperature),* with all gases at a pressure of 1 atm and all solutions at a concentration of 1 mole/liter. All species are taken to be in their most stable state. For substances normally solid at room temperature, the crystal form must often be specified as well. A superscript $^{\circ}$ is used to indicate these standard conditions. Thus,

$$\Delta H_f^{\circ}(H_2O(g)) = -57.796 \text{ kcal/mole} \tag{2-29}$$

means that when 1 mole of water vapor is formed from 1 mole of hydrogen gas and $\frac{1}{2}$ mole of oxygen gas at room temperature (25°C = 298.16°K), 57.796 kcal of heat are given off.

Some representative values of ΔH_f° are collected in Table 2-1. (A much larger collection is given in Appendix B.) An element already in its standard state naturally has a ΔH_f of zero. An element not in its standard state has $\Delta H_f \neq 0$.

Table 2-1 Some Standard Heats of Formation, ΔH_f°(kcal/mole)

Elements		**Compounds**	
$F_2(g)$	0.0	$CO(g)$	−26.4
$Cl_2(g)$	0.0	$CO_2(g)$	−94.1
$Br_2(\ell)$	0.0	$H_2O(\ell)$	−68.3
$Br_2(g)$	+7.4	$H_2O(g)$	−57.8
$I_2(s)$	0.0	$C_2H_6(g)$	−20.2
$I_2(g)$	+14.9	$C_2H_4(g)$	+12.5
$F(g)$	+18.9	$C_2H_2(g)$	+54.2
C(s, graphite)	0.0	$CaO(s)$	−151.9
C(s, diamond)	+0.45	$Ca(OH)_2(s)$	−235.8
$C(g)$	+171.3	$CaCO_3(s)$	−288.4
$O_2(g)$	0.0	$SO_2(g)$	−70.9
$S_8(s)$	0.0	$SO_3(g)$	−94.6

Let's use some of these data to calculate ΔH° for the reaction

$$SO_2(g) + \tfrac{1}{2}O_2(g) \rightleftarrows SO_3(g) \tag{2-30}$$

The enthalpy change associated with this process can easily be

*25°C = 298.16°K.

calculated from $\Delta H_f^\circ(SO_2)$ and $\Delta H_f^\circ(SO_3)$, which refer to the following processes:

$$\tfrac{1}{8}S_8(s) + O_2(g) \rightleftarrows SO_2(g) \qquad \Delta H_f^\circ(SO_2) = -70.9 \text{ kcal/mole} \qquad (2\text{-}31)$$

$$\tfrac{1}{8}S_8(s) + \tfrac{3}{2}O_2(g) \rightleftarrows SO_3(g) \qquad \Delta H_f^\circ(SO_3) = -94.6 \text{ kcal/mole} \qquad (2\text{-}32)$$

Figure 2-9 illustrates the reasoning involved. We want to know ΔH_{I}. However, since H is a state function, ΔH_{I} must be equal to ΔH_{II}, the heat evolved when SO_2 is converted to SO_3 via the two-step process. The first step is the reverse of the heat of formation

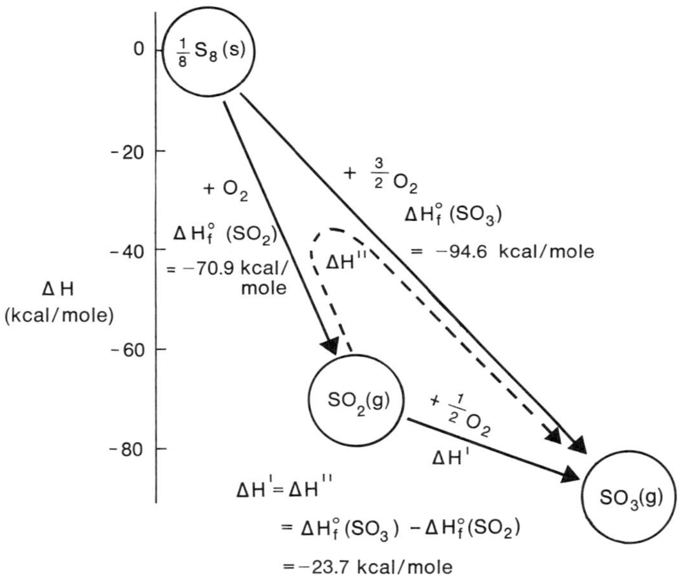

Figure 2-9 Calculation of ΔH for the process $SO_2(g) + \tfrac{1}{2}O_2(g) \rightleftharpoons SO_3(g)$ from standard enthalpies of formation of SO_2 and SO_3.

of SO_2 from the elements, and the second is ΔH_f° for SO_3. Thus, it follows that

$$\Delta H_{\text{I}} = \Delta H_{\text{II}} = \Delta H_f^\circ(SO_3) - \Delta H_f^\circ(SO_2) \qquad (2\text{-}33)$$

This may be verified from (2-31) and (2-32) by reversing (2-31) and algebraically adding them together:

$$SO_2 \rightleftarrows \tfrac{1}{8}S_8 + O_2 \qquad \Delta H_1^\circ = +70.9 = -\Delta H_f^\circ(SO_2) \qquad (2\text{-}34)$$

$$\underline{\tfrac{1}{8}S_8 + \tfrac{3}{2}O_2 \rightleftarrows SO_3} \qquad \underline{\Delta H_2^\circ = -94.6 = \Delta H_f^\circ(SO_3)} \qquad (2\text{-}35)$$

$$SO_2 + \tfrac{1}{2}O_2 \rightleftarrows SO_3 \qquad \Delta H_3^\circ = -23.7 \text{ kcal/mole} \qquad (2\text{-}36)$$

The enthalpy change ΔH_3°, associated with reaction (2-36), is the algebraic sum of the enthalpy changes of the two component steps ΔH_1° and ΔH_2°:

$$\Delta H_3^\circ = \Delta H_1^\circ + \Delta H_2^\circ \qquad (2\text{-}37)$$

We can substitute the two heats of formation for ΔH_1 and ΔH_2:

$$\Delta H_3^\circ = \Delta H_f^\circ(SO_3) - \Delta H_f^\circ(SO_2) \qquad (2\text{-}38)$$

Examination of (2-38) shows that the desired heat of reaction was obtained by taking the difference between the standard heat of formation of the products and the standard heat of formation of the reactants, remembering that the standard heat of formation of O_2, already in the standard state, is zero. This result is a perfectly general one that can be called the *Law of Additivity of Reaction Heats.** The enthalpy change associated with any reaction is given, in general, by

$$\boxed{\Delta H^\circ = \Sigma\ \Delta H_f^\circ(\text{products}) - \Sigma\ \Delta H_f^\circ(\text{reactants})} \qquad (2\text{-}39)$$

where Σ means "sum of."

By measuring and tabulating ΔH_f° for a few hundred compounds, it is possible to calculate ΔH for the thousands of reactions involving these compounds.

(f) HEAT OF COMBUSTION

It is not always possible to measure the heat of formation by the direct combination of elements—indirect methods must then be used. For example, it is experimentally impossible to measure directly the heat of the reaction

$$2C(\text{graphite}) + 2H_2(g) \rightleftarrows C_2H_4(g) \qquad (2\text{-}40)$$

Fortunately the hydrocarbon, ethylene (C_2H_4), can easily be burned in the presence of oxygen to form water and carbon dioxide:

$$C_2H_4(g) + 3O_2(g) \rightleftarrows 2CO_2(g) + 2H_2O(\ell) \qquad (2\text{-}41)$$

The heat associated with the combustion of ethylene is called the *heat of combustion* $\Delta H_{\text{comb.}}$ and is quite easily measured.†

*An old-fashioned and less self-explanatory name is the "Law of Constant Heat Summation."

†In this section heat given off during combustion will be given a negative sign in accord with usual conventions. Unfortunately exothermic heats of combustion are sometimes tabulated as positive quantities.

Knowledge of ΔH_f° for carbon dioxide and liquid water (Table 2-1) makes it possible to determine $\Delta H_f^\circ(C_2H_4)$ by an appropriate combination of the following set of equations:

$$\text{I} \qquad 2CO_2(g) + 2H_2O(\ell) \rightleftarrows C_2H_4(g) + 3O_2(g) \qquad -\Delta H_{\text{comb.}}(C_2H_4)$$

$$\text{II} \qquad 2C(\text{graphite}) + 2O_2(g) \rightleftarrows 2CO_2(g) \qquad 2\Delta H_f^\circ(CO_2)$$

$$\text{III} \qquad 2H_2(g) + O_2(g) \rightleftarrows 2H_2O(\ell) \qquad 2\Delta H_f^\circ(H_2O)$$

Summing these three equations we get

$$2C(\text{graphite}) + 2H_2(g) \rightleftarrows C_2H_4(g)$$

and

$$\boxed{\Delta H_f^\circ(C_2H_4) = 2\Delta H_f^\circ(CO_2) + 2\Delta H_f^\circ(H_2O) - \Delta H_{\text{comb.}}(C_2H_4)} \qquad (2\text{-}42)$$

A flow chart showing our alternate route between graphite, hydrogen gas and ethylene is given in Figure 2-10. Notice that reactions II and III are also combustion reactions. The heats of formation of carbon dioxide and water are identical with the heats of combustion of hydrogen gas and graphite.

(g) BOND ENERGIES: DIATOMICS

When a chemical reaction occurs, reactant molecules are broken up and product molecules are formed. The energy changes that accompany chemical reactions result from changes in the relative positions of atoms. The potential energy associated with each chemical bond is unique to that bond. It is easy to see the effect of these changes in potential energy when we look in detail at a reaction involving diatomic molecules. Consider the formation of hydrogen fluoride gas from hydrogen gas and fluorine gas:

$$H_2(g) + F_2(g) \rightleftarrows 2HF(g) \qquad (2\text{-}43)$$

The enthalpy change associated with this reaction, as written, is twice the enthalpy of formation of HF (Appendix B):

$$\Delta H = 2\Delta H_f^\circ(HF) = -129.6 \text{ kcal/mole} \qquad (2\text{-}44)$$

Remember that ΔH_f° is defined as the enthalpy change associated with the formation of *one mole* of product. Equation (2-43) involves the formation of *two* moles of HF.

We might think of the formation of HF in the following series of steps, which are illustrated in the flow chart of Figure 2-11.

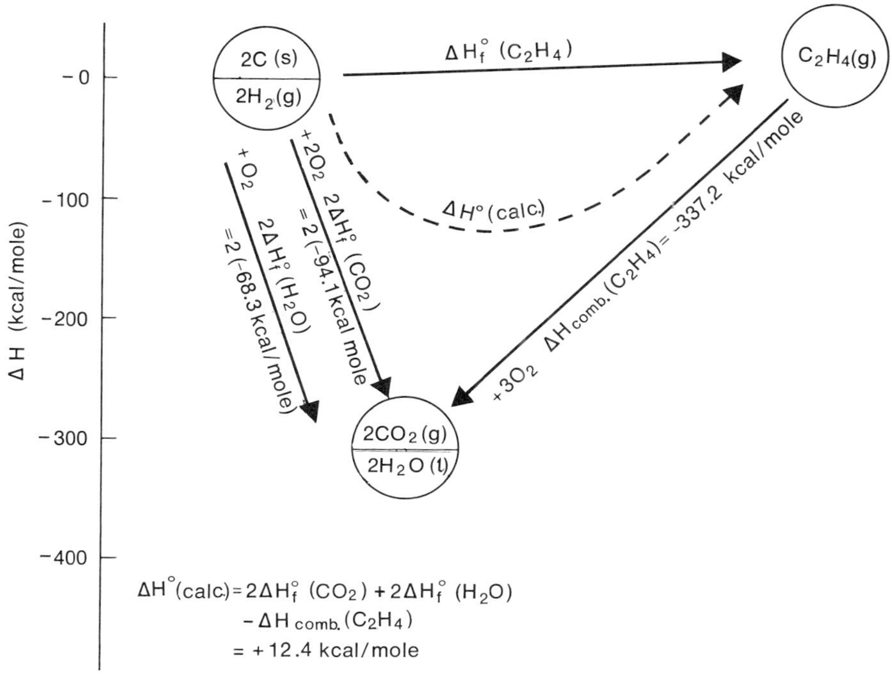

Figure 2-10 Calculation of ΔH_f° (ethylene) from the heats of combustion of ethylene, and ΔH_f° of H_2O and CO_2.

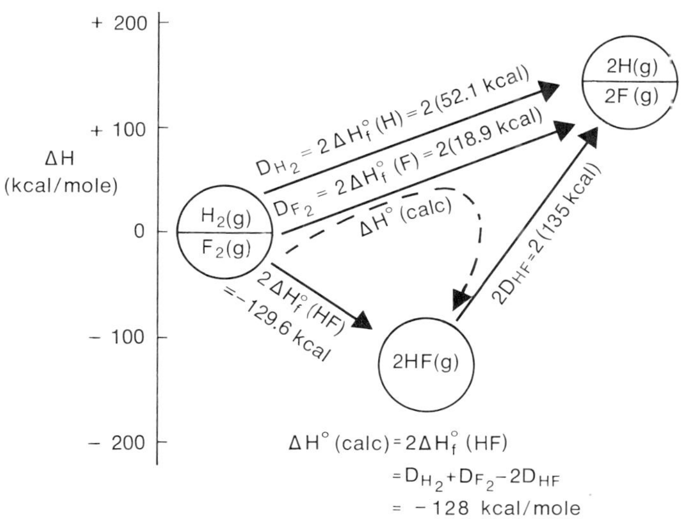

Figure 2-11 Relationship between bond dissociation energies and enthalpies of formation for the process $H_2(g) + F_2(g) = 2HF(g)$.

First, we must separate hydrogen and fluorine atoms from the molecules and let them recombine as HF.

I	$H_2(g) \rightleftarrows 2H(g)$	$\Delta H^{I} = 2\Delta H_f^\circ(H) = D_{H_2}$
II	$F_2(g) \rightleftarrows 2F(g)$	$\Delta H^{II} = 2\Delta H_f^\circ(F) = D_{F_2}$
III	$2H(g) + 2F(g) \rightleftarrows 2HF(g)$	$\Delta H^{III} = -2D_{HF}$

The enthalpy change associated with the net reaction (2-43) is given by the sum of the enthalpy changes of the three steps:

$$\Delta H = 2\Delta H_f^\circ(HF) = \Delta H^{I} + \Delta H^{II} + \Delta H^{III} \qquad (2\text{-}45)$$

Equation I represents the formation of hydrogen atoms from molecular hydrogen, the standard state. This involves breaking the H_2 chemical bond. The energy required to do this is equivalent to the potential energy stored in this bond and is generally called the *bond dissociation energy** and given the symbol D_{H_2}. The same is true for II which represents the breaking of the F—F bond in F_2 and requires an amount of energy equal to the bond dissociation energy of F_2. In III, the combination of hydrogen atoms and fluorine atoms gives off an amount of energy related to the potential energy of the HF bond. In fact, the reverse of this process,

$$\text{IIIa} \qquad 2HF = 2H(g) + 2F(g) \qquad (2\text{-}46)$$

is just an expression for the dissociation of two molecules of HF. The enthalpy change associated with III is then twice the negative (because we are considering the reverse reaction) of this bond dissociation energy of HF.

If we start with one molecule each of H_2 and F_2, we must first put in an amount of energy equivalent to the potential energy of the F—F and H—H bonds. Formation of two molecules of HF gives us back the potential energy of the HF bond for each molecule formed. We can thus rewrite (2-45) in terms of the bond dissociation energies:

$$\Delta H = 2\Delta H_f^\circ(HF) = D_{H_2} + D_{F_2} - 2D_{HF} \qquad (2\text{-}47)$$

(h) BOND ENERGIES: POLYATOMICS

The energy of a particular bond in a polyatomic molecule is not as easy to pin down as the bond dissociation energy of a diatomic

*It is customary to speak of bond dissociation *energies* even though the quantity we are actually measuring in this example is the bond dissociation *enthalpy*. A detailed consideration of the relation between ΔE and ΔH (taken up later in this chapter) shows that the two are nearly equal for processes such as these.

molecule. For example, the energy (enthalpy) required to break a carbon—hydrogen bond in some organic compounds is tabulated in Table 2-2. The range is considerable, about 10 percent among the examples given.

Table 2-2 Some Carbon—Hydrogen Bond Energies

Compound	D_{C-H} (kcal/mole)
CH_4	101
CH_3CH_3	96
$CH_3CH_2CH_3$	100(CH_3)
$CH_3CH_2CH_3$	94(CH_2)
C_6H_6	102
$CHCl_3$	90
CH_3Br	99

It would be useful, however, if some useful values of these bond energies were at hand. For example, there are over a million organic compounds which involve, at most, 25 different bonds. In order to calculate the enthalpy changes associated with the countless reactions these molecules undergo, it would be necessary to tabulate ΔH_f° for every one of these. This is a formidable task. However, if the heat of a given reaction could be related to the energies involved in making and breaking bonds, a relatively small table of bond energies would suffice. For this purpose, collections of *average* bond energies have been prepared, which give representative bond energies based on a large selection of compounds in which that bond appears. A short list of these is given in Table 2-3. Calculations performed with these values give only approximate ΔH's, but they provide useful indications when experimental values are not available.

For example, let's consider the hydrogenation of carbon tetrachloride to chloroform:

$$CCl_4(g) + H_2(g) \rightleftarrows CHCl_3(g) + HCl(g) \tag{2-48}$$

A block diagram for this process is given in Figure 2-12. The first step is the breaking of a C—Cl bond and an H—H bond. This is followed by the formation of a C—H bond and an H—Cl bond. The enthalpy of this process is given approximately by

$$\begin{aligned} \Delta H &= \underset{\text{energy req.}}{(D_{C-Cl} + D_{H-H})} - \underset{\text{energy back}}{(D_{C-H} + D_{H-Cl})} \\ &= (81 + 104) - (99 + 103) \\ &= -17 \text{ kcal/mole} \end{aligned} \tag{2-49}$$

The true ΔH obtained from measured heats of formation is −22.1 kcal/mole. So, while the answer obtained using the average

Table 2-3 Some Average Bond Energies

Bond	Average Energy (kcal/mole)
C—C	82.6
C═C	145.8
C≡C	199.6
H—H	104.2
O═O	119.1
N≡N	225.8
F—F	37
Cl—Cl	57.9
Br—Br	46.1
I—I	36.1
C—H	98.8
C—F	116
C—Cl	81
C—Br	68
C—I	51
C—O	85.5
C═O (aldehydes & ketones)	178
C═O (CO_2)	192.1
O—H	110.6
H—F	135
H—Cl	103.1
H—Br	87.4
H—I	71.4
N—H	93.4
N—F	65
N—Cl	46

bond energies is only approximate, it is fairly close to the correct answer, and gives us a good indication of the magnitude of the energy changes involved.

(i) PHASE CHANGES: LIQUID TO GAS

If a liquid is slowly heated, its temperature gradually rises until the boiling point is reached. The temperature then stays constant until all the liquid has been converted into vapor. The heat required to vaporize one mole of a liquid at its boiling point is called the heat or enthalpy of vaporization and is given the symbol $\Delta H_{\text{vap.}}$. This quantity is related to the energy required to pull a molecule out of the liquid and put it into the gas phase. Because there is a change in volume, some extra work has to be done by the escaping gas molecules to push back the atmosphere. Thus the measured quantity $\Delta H_{\text{vap.}}$ will be greater than the energy $\Delta E_{\text{vap.}}$ required to separate a molecule from the liquid.

It is quite easy to estimate the difference between $\Delta H_{\text{vap.}}$ and $\Delta E_{\text{vap.}}$. Let's do this for $H_2O(\ell) \rightleftarrows H_2O(g)$. For a constant pres-

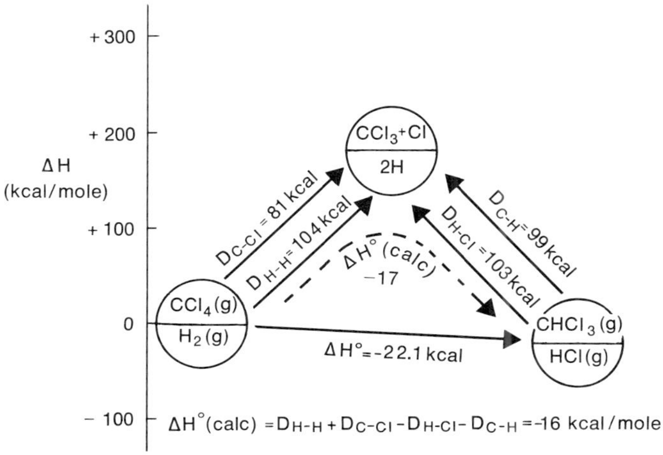

Figure 2-12 Bond energy route for estimation of ΔH. The value so derived (−17 kcal/mole) is a reasonable approximate value for ΔH°, measured to be −22.1 kcal/mole.

sure process, we can write

$$\Delta H = \Delta E + P\Delta V \tag{2-50}$$

and solve for ΔE,

$$\Delta E_{\text{vap.}} = \Delta H_{\text{vap.}} - P\Delta V \tag{2-51}$$

The volume change is given by

$$\Delta V = V_{\text{vapor}} - V_{\text{liquid}} \tag{2-52}$$

and it is almost always a good approximation to assume that the gas volume of a given mass of a substance is much larger than the liquid volume of that same mass. If so, we can write

$$\Delta V \sim V_{\text{vapor}}$$

Let's check this assumption for water before going any further. The volumes occupied by one gram of liquid water and water vapor at 100°C are 1.04 ml and 1.68 liter, respectively. So we can see immediately that there is at least a one-thousand-fold increase in volume on vaporization. For one mole of water (18 g) we have

$$\begin{aligned}\Delta V &= V_{\text{vapor}} - V_{\text{liquid}} \\ &= 30.2 - 0.019 \\ &= 30.18 \text{ liters } (\approx 30.2 \text{ liters}) \\ &\cong V_{\text{vapor}}\end{aligned} \tag{2-53}$$

If we assume that water vapor behaves as an ideal gas, we can write

$$PV_{\text{vapor}} = n_{\text{vapor}}RT \qquad (2\text{-}54)$$

Since we are worrying about ΔH and ΔE on a one-mole basis,

$$n_{\text{vapor}} = 1 \text{ mole} \qquad (2\text{-}55)$$

and the PV product is easily calculated:

$$\begin{aligned} PV_{\text{vapor}} &= RT \\ &= (1.987 \text{ cal/mole degK}) \cdot (373^\circ\text{K}) \\ &= 741 \text{ cal/mole} \end{aligned} \qquad (2\text{-}56)$$

The observed enthalpy of vaporization of water at its boiling point is

$$\Delta H_{\text{vap.}}\ (H_2O) = 9{,}820 \text{ cal/mole}$$

Thus the energy required for the vaporization process is

$$\begin{aligned} \Delta E_{\text{vap.}} &= 9{,}820 - 741 \text{ cal/mole} \\ &= 9{,}080 \text{ cal/mole} \end{aligned}$$

Thus only a relatively small fraction, less than 10 percent, of $\Delta H_{\text{vap.}}$ represents energy used to push back the atmosphere. We can feel quite relaxed, then, about using the easily determined quantity $\Delta H_{\text{vap.}}$ as an indication of the effort required to evaporate a liquid.

(j) PHASE CHANGES: SOLID TO LIQUID

Energy is required to release molecules from the liquid into the gas phase. Energy is also required to convert a solid into liquid form. The enthalpy change associated with melting is called the enthalpy of fusion $\Delta H_{\text{fus.}}$. Once again it is ΔH, rather than ΔE, which is conveniently measured. In this case, however, the difference between them is indeed negligible. The expression for ΔE is

$$\begin{aligned} \Delta E_{\text{fus.}} &= \Delta H_{\text{fus.}} - P\Delta V \\ &= \Delta H_{\text{fus.}} - P(V_{\text{liquid}} - V_{\text{solid}}) \end{aligned} \qquad (2\text{-}57)$$

The volume occupied by a mole of liquid is, in general, very close to the volume occupied by a mole of the same substance in the solid state. It is, therefore, a very good assumption that

$$\Delta V_{\text{solid} \rightarrow \text{liquid}} = 0 \qquad (2\text{-}58)$$

so that ΔH and ΔE are sensibly equal.

It is interesting to compare the processes of fusion and vaporization—Table 2-4 contrasts $\Delta H_{\text{fus.}}$ and $\Delta H_{\text{vap.}}$ for a number of substances. In every case, $\Delta H_{\text{vap.}}$ is much larger than $\Delta H_{\text{fus.}}$. The reasons are rather clear.

Table 2-4 Enthalpy of Fusion and Vaporization

Substance	$\Delta H_{\text{fus.}}$ ***(kcal/mole)***	$\Delta H_{\text{vap.}}$ ***(kcal/mole)***
HCl	0.48	3.9
HBr	0.58	4.2
H_2S	0.59	4.5
NH_3	1.35	5.6
CCl_4	0.60	7.3
C_6H_6	2.35	7.4
H_2O	1.44	9.8
H_2O_2	2.92	10.3

The change from solid to liquid requires that the tight and regular arrangement of molecules in the solid be loosened enough to permit the freedom of movement characteristic of fluids. The similarity of molar volumes tells us, however, that molecules in a liquid are still close together; hence held quite tightly. In the gas phase, on the other hand, molecules are so far apart that there are only very small forces between them (none in an ideal gas). Much more energy, then, has to be expended to remove a molecule from the liquid into the gas phase, than to just loosen up the solid into a liquid.

2-4 Calorimetry

Until now we have blithely talked about thermochemistry without paying any attention to the experimental details of the measurements involved. In this concluding section, we will discuss the nature of the absorption of heat, and describe some of the experimental methods used.

(a) HEAT CAPACITY: MOLECULAR ORIGINS

Consider a bulb containing one mole of a monatomic gas at room temperature. Suppose the bulb is warmed in a constant-temperature bath. The absorption of heat by the gas and the bulb results in an increase in the average translational kinetic energy of the molecules. This increase in translational energy is manifested as a temperature rise. If the bulb is cooled, the molecules move more and more slowly until, at the absolute zero, there is no translational motion at all (the gas has by this time condensed to a solid).

Of course a polyatomic molecule has vibrational and rotational kinetic energy, as well as translational. When such a molecule is heated, some of the added energy goes towards increasing vibration and rotation, so that it will take more energy to achieve a given temperature rise in a polyatomic gas than in a monatomic gas. The *heat capacity* of a substance is defined as the amount of heat, in calories, necessary to effect a one-degree rise in temperature. In thermodynamics we will primarily be interested in processes carried out at constant pressure and will, as usual, refer our measurements to the standard quantity, the mole. Therefore, we will deal with the *molar heat capacity at constant pressure,* C_p. This is the heat in calories necessary to raise the temperature of one mole of substance one degree at constant pressure. Thus, the heat required for a temperature rise ΔT at constant pressure is given by

$$\Delta H = q_P = C_p \Delta T \tag{2-59}$$

$$\Delta T = (T_{\text{final}} - T_{\text{initial}})$$

where T is the absolute temperature in degrees K, and C_p is the molar heat capacity at constant pressure.

It should be clear from what has been said, that the molecular size and molecular geometry should have an important influence on the heat capacity. A large molecule with lots of atoms has many more ways of absorbing added energy in vibration and rotation than does, say, carbon monoxide, a diatomic gas. Some representative values of C_p are given in Table 2-5.

The upper portion of this table shows the dependence of C_p on molecular complexity. The monatomic gases all have identical heat capacities—the natural result of the fact that all the absorbed energy must go into translational motion. (Only at very high tem-

Table 2-5 Molar Heat Capacities at Constant Pressure (cal/mole degK)

Monatomic Gases		**Diatomic Gases**		**Triatomic Gases**	
He	4.9	CO	7.0	H_2O	8.0
Ne	4.9	N_2	7.0	D_2O	8.2
Ar	4.9	F_2	7.5	CO_2	8.9
Xe	4.9	Cl_2	8.1	CS_2	10.9
C	4.9				

Tetratomic Gases		**Polyatomic Gases**	
NH_3	8.5	SF_6	29.0
H_2CO	8.5	UF_6	31.0
Cl_2CO	14.5	C_2H_6	12.6
		CH_3NHCH_3	16.9

Pentatomic Gases			
CH_4 (MW = 16)	8.4	CH_3Cl (MW = 51)	9.8
CF_4 (MW = 88)	14.6	CH_2Cl_2 (MW = 85)	12.3
CCl_4 (MW = 154)	19.9	$CHCl_3$ (MW = 120)	15.7
CI_4 (MW = 520)	22.9		

peratures do the effects of electronic kinetic energy become noticeable.) As the molecules get more and more complex, their heat capacities rise steadily. The extra energy goes into vibrational and rotational motion.

The lower half of the table illustrates another important point. For these similar pentatomic molecules, the heat capacity also increases as more and more heavy atoms are added. This is because the vibrations and rotations of molecules are *quantized* (the same effect that results in the line spectrum of the hydrogen atom). The heavier atoms in a molecule vibrate more slowly, so their vibrational energy levels are closer together. Hence, more vibrations can be excited at a given temperature in a "heavy-atom" molecule like CI_4 than in a "light-atom" molecule like CH_4. The accumulation of this effect can be seen most vividly in the sequence CH_4, CH_3Cl, CH_2Cl_2, $CHCl_3$, CCl_4.

(b) THE ICE CALORIMETER

The device illustrated in Figure 2-13 is known as an ice calorimeter and is commonly used for the measurement of the heat capacity of solids and liquids, or reaction heats. The operation of

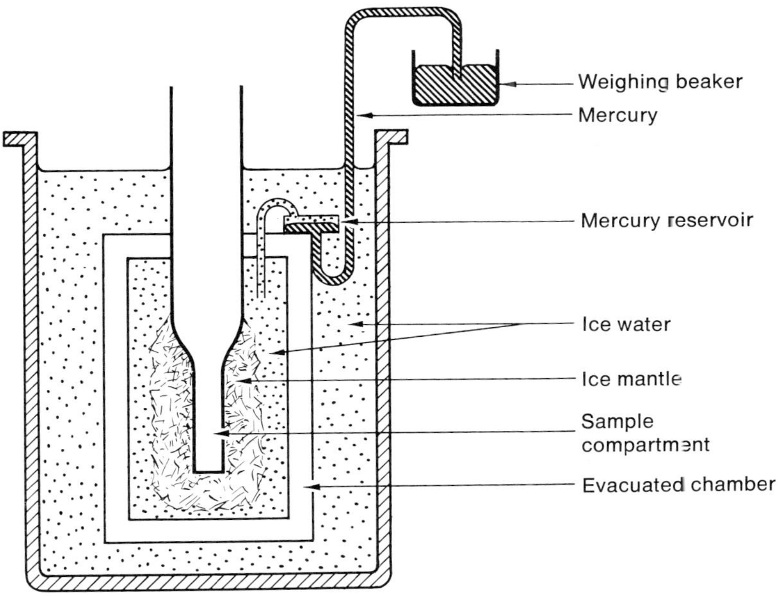

Figure 2-13 Ice calorimeter. A hot sample placed in the sample compartment melts some of the ice mantle. The decrease in volume sucks mercury from the weighing beaker into the mercury reservoir. (This is a simplified drawing of an ice calorimeter described by D. C. Ginnings and R. C. Corruccini, Journal of Research of the National Bureau of Standards, **38,** *p. 583, 1947.)*

this calorimeter depends on the fact that the volume of a given mass of ice is *greater* than the volume of an equivalent mass of liquid water. At the beginning of the experiment, a mantle of ice is frozen around the sample tube by placing dry ice (solid CO_2) inside the tube. The mercury in the weighing beaker is accurately weighed. Then the hot sample is dropped into the container and allowed to cool to 0°C. As the sample cools, the ice melts and the total volume of ice and water decreases, thereby sucking mercury from the weighing beaker into the reservoir. When temperature equilibrium has been reached, the beaker is reweighed. The change in volume is proportional to the amount of ice melted and can be calculated from the weight of mercury pulled from the beaker. Heat losses to the surroundings are minimized by placing the entire apparatus in a large vessel containing an ice–water mixture.

The apparatus is calibrated by heating the sample tube with a heating coil to which an accurately known quantity of electrical power is supplied.

In a typical calibration experiment with an ice calorimeter, 61.4 grams of mercury were withdrawn from the weighing beaker. The heater was run for 70 minutes with an average current of 0.60 amperes and an average voltage of 66 volts. We want to find the calibration constant K_g of the calorimeter in calories per gram of mercury:

$$\text{electrical energy (joules)} = Vit \tag{2-60}$$

where V is volts, i is amperes, and t is time in seconds (1 cal = 4.2 joules). Therefore,

$$E_{\text{elect.}} = \frac{(66\text{V}) \cdot (0.60\text{ A}) \cdot (420\text{ sec})}{(4.2\text{ joules/cal})} = 3.96 \times 10^3\text{ cal}$$

Now we can find the heat dissipated per gram of mercury withdrawn from the weighing beaker. We will call this quantity K_g, the calibration constant of the apparatus.

$$K_g = \frac{(3.96 \times 10^3\text{ cal})}{(61.4\text{ g Hg})} = 64.6\text{ cal/g Hg} \tag{2-61}$$

Thus a sample must deliver 64.6 cal of heat to change the weight of the weighing beaker by one gram.

We can find C_p for carbon tetrachloride from the following experimental results: 103 ml of CCl_4 (3.0 moles) at 50°C (323°K) were placed in the ice calorimeter calibrated in the previous example. A total of 7.3 grams of mercury were sucked from the weighing beaker.

The heat lost by sample was

$$
\begin{aligned}
\Delta H &= K_g \cdot \text{weight} \\
&= (64.6 \text{ cal/g Hg}) \cdot (7.3 \text{ g Hg}) \\
&= 472 \text{ cal}
\end{aligned} \tag{2-62}
$$

The heat capacity for one mole is given by (2-59):

$$\Delta H = C_p \Delta T$$

When n moles are involved

$$\Delta H = nC_p \Delta T$$

Solving for C_p,

$$
\begin{aligned}
C_p &= \frac{\Delta H}{n\Delta T} \\
&= \frac{(472 \text{ cal})}{(3 \text{ moles}) \cdot (50°\text{K})} \\
C_p &= 31.5 \text{ cal/mole degK}
\end{aligned}
$$

(c) THE BOMB CALORIMETER: HEATS OF COMBUSTION

The ice calorimeter is useful for measuring heat capacities and enthalpy changes of reactions occurring in solution. Another type of calorimeter in wide use is the bomb calorimeter, illustrated in Figure 2-14. This calorimeter is used for measuring heats of combustion. The sample is placed in a small cup inside the tightly sealed bomb, and a large pressure of O_2 is admitted. The bomb is placed in a vessel and covered with water. This part of the apparatus is then placed in a large insulated vessel to minimize heat losses.* The sample is ignited by means of an electrical pulse through the ignition wires. The temperature rise of the water in the calorimeter is measured. Of course the apparatus has to be calibrated. This is usually done by igniting a sample with an accurately known heat of combustion. This allows the total heat capacity of the apparatus and water to be determined.

Any process carried out inside such a tightly closed bomb is obviously occurring at constant volume. We are measuring, then, the heat of combustion at constant volume, which is equivalent to the change in energy of the system:

$$q_V = \Delta E \tag{2-63}$$

*The apparatus shown in Figure 2-14 is a simplified version of that normally used. A conventional bomb calorimeter keeps heat losses to a minimum by keeping the outer jacket of the apparatus at the same temperature as the calorimeter. This is done either by electric heating or by running warm water through the jacket.

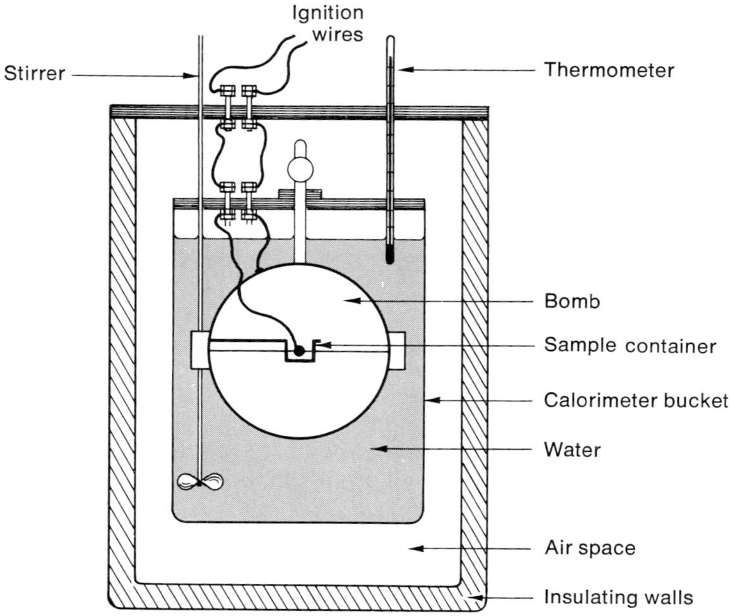

Figure 2-14 A simplified bomb calorimeter.

However, we want ΔH, the enthalpy change, not ΔE, if we are to keep all our thermochemical results consistent. Let's consider the difference between these two quantities. From the definition of enthalpy

$$H = E + PV \qquad (2\text{-}64)$$

we find that the enthalpy *change* is given by

$$\Delta H = \Delta E + \Delta(PV) \qquad (2\text{-}65)$$

If the products *and* reactants were all solids and liquids, the term $\Delta(PV)$ would be negligibly small and we could assume that

$$\Delta H = \Delta E \quad \text{(solids and liquids)} \qquad (2\text{-}66)$$

If gaseous atoms are among the products or reactants, however, this is not the case. Suppose that any gaseous species involved behave as ideal gases, and thus, follow the ideal gas law

$$PV = nRT \qquad (2\text{-}67)$$

We can then find $\Delta(PV)$ in terms of the other quantities as follows:

$$\begin{aligned}\Delta(PV) &= \Delta n(RT) \\ &= (n_2 - n_1)RT\end{aligned} \qquad (2\text{-}68)$$

where n_2 is the number of moles of gaseous products

n_1 is the number of moles of gaseous reactants

R is the gas constant (1.987 cal/mole degK)

T is the average temperature (absolute) during measurement

The correction necessary to convert a measured ΔE into a ΔH depends, then, on the *difference* between the number of moles of gases appearing as products and reactants. As we shall see in the following example, the correction is usually not very large, even when several moles of gases are involved. Hence the use of an average temperature introduces a negligible error.

In a typical experiment* using a bomb calorimeter, a sample of biphenyl $(C_6H_5)_2$ weighing 0.526 g was ignited in an excess of oxygen gas. The observed temperature rise was 1.91 degrees. In a separate calibration experiment, a sample of benzoic acid, C_6H_5COOH, weighing 0.825 g was ignited and produced a temperature rise of 1.94 degrees. The energy of combustion of benzoic acid (BA) is accurately known

$$\Delta E_{\text{comb.}}(\text{benzoic acid}) = \Delta E_{\text{comb.}}(\text{BA}) = -771 \text{ kcal/mole} \quad (2\text{-}69)$$

The water in the calorimeter was initially at room temperature 25°C (298°K). We can calculate $\Delta E_{\text{comb.}}$ and $\Delta H_{\text{comb.}}$ for biphenyl. The total heat capacity of the calorimeter and water must be calculated first from the known heat delivered by the combustion of benzoic acid:†

$$-\Delta E_{\text{comb.}}(\text{BA}) = C(\text{calorimeter}) \cdot \Delta T$$

$$C(\text{calorimeter}) = \frac{(771 \text{ kcal/mole}) \cdot (0.825 \text{ g})}{(1.94 \text{ degK}) \cdot (123 \text{ g/mole})}$$

$$= 2.66 \text{ kcal/degK}$$

Using this result, we can calculate the energy of combustion of biphenyl (BP):

$$-\Delta E_{\text{comb.}}(\text{BP}) = C\ (\text{calorimeter}) \cdot \Delta T$$

$$\Delta E_{\text{comb.}}(\text{BP}) = -\frac{(2.66 \text{ kcal/degK}) \cdot (1.91 \text{ degK}) \cdot (154 \text{ g/mole})}{0.526 \text{ g}}$$

$$= -1490 \text{ kcal/mole}$$

*These results were obtained in a senior physical chemistry laboratory at the University of British Columbia.

†Heat given off by the combustion process is absorbed by the calorimeter. The heat absorbed is thus the negative of the heat of combustion, $-\Delta E_{\text{comb.}}$.

Now we must convert $\Delta E_{\text{comb.}}$ to $\Delta H_{\text{comb.}}$. The chemical equation is

$$(C_6H_5)_2(s) + \tfrac{29}{2}O_2(g) \rightleftarrows 12CO_2(g) + 5H_2O(\ell) \qquad (2\text{-}70)$$

The change in moles of gaseous substances is

$$\begin{aligned}\Delta n \text{ (gases)} &= (n_{\text{products}} - n_{\text{reactants}}) \\ &= 17 - \tfrac{29}{2} \\ &= \tfrac{5}{2}\end{aligned} \qquad (2\text{-}71)$$

Substituting this into (2-68) and (2-65) gives us

$$\begin{aligned}\Delta H_{\text{comb.}} &= \Delta E_{\text{comb.}} + \tfrac{5}{2}(1.987 \text{ cal/mole degK}) \cdot (298°\text{K}) \\ &= (-1490 + 1.4) \text{ kcal/mole} \\ &= -1489 \text{ kcal/mole}\end{aligned}$$

Some experimentally measured heats of combustion are given in Table 2-6.

*Table 2-6 Enthalpies of Combustion of Organic Compounds**

Name	Formula	$\Delta H_{\text{comb.}}$ at 298°K (kcal/mole)†
Methane	$CH_4(g)$	−212.8
Ethane	$C_2H_6(g)$	−372.8
Propane	$C_3H_8(g)$	−530.6
Ethylene	$C_2H_4(g)$	−337.2
Benzene	$C_6H_6(\ell)$	−781.0
Methanol	$CH_3OH(\ell)$	−173.6
Ethanol	$C_2H_5OH(\ell)$	−326.7
Dimethyl ether	$(CH_3)_2O(g)$	−347.8
Acetic acid	$CH_3COOH(\ell)$	−209.4
Glucose	$C_6H_{12}O_6(s)$	−673
Sucrose	$C_{12}H_{22}O_{11}(s)$	−1348.9

*The final products are $CO_2(g)$ and $H_2O(\ell)$; $\Delta H_f^\circ(CO_2(g)) = -94.1$ kcal; $\Delta H_f^\circ(H_2O(\ell)) = -68.3$ kcal/mole.

†Tables of combustion measurements are often reproduced with the opposite sign convention so that heats of combustion are listed as positive quantities.

(d) CALORIMETRIC DETERMINATION OF $\Delta H_{\text{vap.}}$

The energy (enthalpy) required to vaporize a mole of a pure liquid can be determined by supplying a measured amount of electric power. A suitable apparatus is illustrated in Figure 2-15. The liquid is placed in an insulated flask and an accurately known amount of electrical energy is supplied through the heating coil. The vapor is condensed in the water-cooled condenser and collected in the receiving flask.

A single measurement performed in this way would not be too

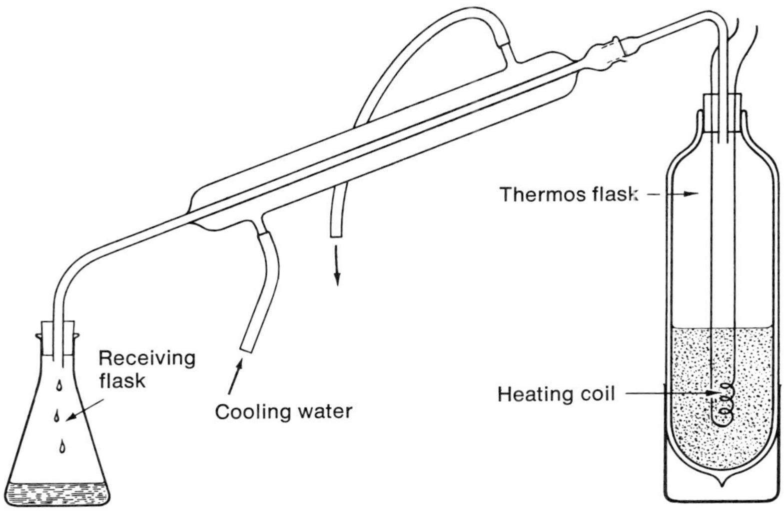

Figure 2-15 Apparatus for the calorimetric determination of $\Delta H_{vap.}$

satisfactory because of the heat losses to the apparatus. If measurements are made at several rates of evaporation, an accurate value of $\Delta H_{\mathrm{vap.}}$ can be obtained.

The rate at which electrical energy is supplied in watts (joules per second) is given by the product:

$$\text{power} = Vi$$

where V is volts and i is amperes.

This is balanced by the evaporation process which occurs at the rate,

$$\text{rate of evaporation} = \frac{n\Delta H_{\mathrm{vap.}}}{t}$$

where n is the number of moles
t is the time electrical power is supplied

section 2-4
calorimetry

and by the losses to the apparatus, which we shall call h. Thus we can combine these factors to give us

$$\text{power} = Vi = \frac{n\Delta H_{\mathrm{vap.}}}{t} + h$$

If we perform experiments at several rates of evaporation (n/t), a plot of this rate against electrical power should give a linear plot of slope $\Delta H_{\mathrm{vap.}}$ and intercept h.*

*This equation has the form of the general linear equation $y = ax + b$, in which a is the slope and b the intercept in the y axis.

Using the apparatus described above,* the following results were obtained for chloroform, $CHCl_3$. (1 cal = 4.18 joules)

	power, Vi (joules/sec)	rate of distillation, n/t (mole/sec)
I	8.75	2.15×10^{-4}
II	6.84	1.47×10^{-4}

We can calculate $\Delta H_{\text{vap.}}$ for chloroform using only these two pieces of data. (A real determination would use several experimental rates and find $\Delta H_{\text{vap.}}$ from the slope of the graph described above.)

We can write

$$(Vi)_{\text{I}} = \Delta H_{\text{vap.}}(n/t)_1 + h \tag{2-72}$$

and

$$(Vi)_{\text{II}} = \Delta H_{\text{vap.}}(n/t)_2 + h \tag{2-73}$$

Subtracting equation (2-72) from equation (2-73), we have

$$\Delta(Vi) = \Delta H_{\text{vap.}}\Delta(n/t)$$

Thus,

$$\begin{aligned} \Delta H_{\text{vap.}} &= \frac{\Delta(Vi)}{\Delta(n/t)} \\ &= \frac{(8.75 - 6.84 \text{ joules/sec})}{[(2.15 - 1.47) \times 10^{-4} \text{ moles/sec}] \cdot (4.18 \text{ joules/cal})} \\ &= 6.72 \text{ kcal/mole} \end{aligned}$$

2-5 Energy and spontaneity—Why do things roll downhill?

In Figure 2-2, we saw a runaway car plummet down the hill. The brake having slipped, the car spontaneously began its merry trip. Figure 2-16 shows where it probably completed its thoughtless journey. At the beginning, atop the hill, the car held potential energy. As it neared the bottom, this energy changed into kinetic energy—to the dismay of casual pedestrians! Then everything ended in chaos (see Fig. 2-13).

(a) ROLLING DOWNHILL—A MODEL FOR CHEMISTRY?

Is there a model here concerning chemical events? Back in 1860, Berthelot thought there was, when he postulated that all chemical

*These data were obtained in a first-year chemistry lab at the University of British Columbia.

Figure 2-16 End of a joyride.

changes tend to liberate heat. Having carefully measured the heat released in dozens and dozens of exothermic reactions he became so preoccupied with heat release that he decided it pointed in the direction of spontaneous change. He decided that chemical reactions are like the car in Figure 2-2: reactions seem to "roll downhill" on a chemical energy landscape.

This was a strange conclusion for this brilliant scientist to reach. He had ample evidence at hand that contradicted this simple view. Many familiar changes absorb heat as they occur—salt dissolving in water is one example, and carbon dioxide expanding from a high-pressure cylinder is another. Both changes occur quite spontaneously but obviously without releasing heat. Furthermore, chemists of the time knew of many reactions which, though exothermic, proceed to an equilibrium mixture in which some reactants remain unconsumed. Nevertheless, Berthelot's idea received a certain acceptance. After all, strongly exothermic reactions are often the most difficult to restrain, and the fact that balls do roll downhill seemed to offer some connection. To explore the idea, let's consider a laboratory version of the events of Figure 2-2.

Figure 2-17 shows a steel ball-bearing on a polished metal surface, position (*a*). Since it is on a slope, it begins to roll down the hill, picking up speed as it goes. By the time it is half-way down, position (*b*), it has picked up considerable speed. Much of its potential energy (energy of position) has been converted into kinetic energy (energy of motion). But the ball is still on the hillside, so it accelerates still more. By the time it reaches the bottom of the hill, position (*c*), all of the potential energy is expended and converted into kinetic energy. The ball has spontaneously rolled downhill.

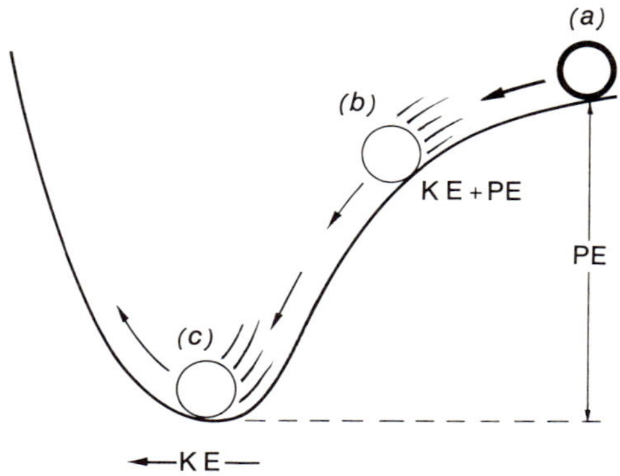

Figure 2-17 A ball-bearing placed on a polished metal hillside rolls downhill.

Now wait a second—see Figure 2-18! Shortly after the ball reaches the bottom of the hill, it starts rolling *up* the hill on the other side. At (*d*) its velocity is reduced, but now the ball again possesses some potential energy. By the time it reaches (*e*), the energy of motion is gone and the ball again possesses only potential energy. So *the ball has now spontaneously rolled uphill.* Apparently it had just as strong a tendency to roll uphill, once it gained a good head of steam, as it had to roll downhill when it was parked on the hillside.

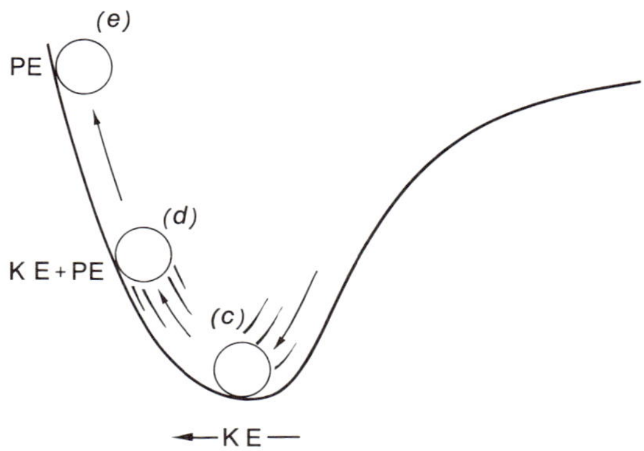

Figure 2-18 A ball-bearing can roll uphill, too.

Figure 2-19 shows the track of the ball when just left to roll back and forth on the hill. Each time it completes a tour, down on the right, up on the left, back down on the left and up on the right, it comes up a little shy of its starting point. As time goes on, the

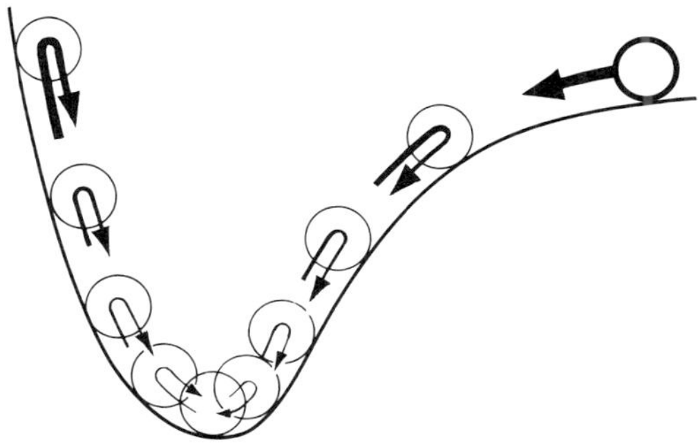

Figure 2-19 Gradually friction spoils the fun.

amplitude of its motion decreases more and more. Finally, it ends up at the bottom of the hill, at rest.

Where, now, is its energy? What happened to the tendency of the ball to roll back and forth; first down, then up? The answer is that friction took a hand. No matter how carefully we polish the hillside, a little friction remains, so the ball rubs against the surface and produces heat. Gradually all the energy gets converted into heat and, as it does, the ball reaches the bottom of the hill.

Recapitulating, we find that when the ball possesses potential energy, it tends to roll downhill. This merely generates kinetic energy, and then the ball tends to roll uphill. The energy of position is converted to directed energy of motion. What finally gets the ball to the bottom of the valley is the dissipation of this directed energy of motion into the random motion of heat.

(b) ATOMS REACTING—ROLLING DOWNHILL?

Figure 2-20 shows a plot of how the energy changes as two iodine atoms approach each other. The horizontal scale shows the internuclear distance, and the vertical scale shows the energy of the electrons as they sense the positive and negative charges on the other atom. The energy drops as they approach until it reaches a minimum at R_0, the equilibrium bond length for I_2.

However, the iodine atoms don't stop at R_0. They are accelerated as they approach each other, so they proceed right through the equilibrium distance and up the other side. On reaching point (*e*), the energy of the iodine atoms returns completely to potential energy. Then they turn around and come back down the hill, through the R_0 position and out again, flying apart. The iodine atoms fail to react!

What they need, according to Figure 2-19, is some friction. And this is true. The friction must be provided by collisions with other molecules, and it must be available in the short time before the

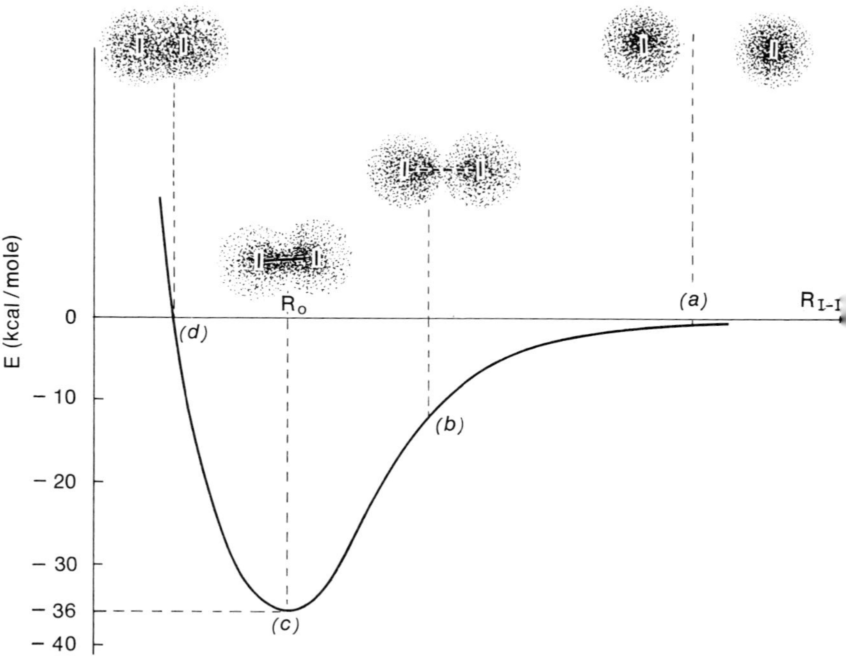

Figure 2-20 Energy of two iodine atoms as they collide.

iodine atoms separate again. If the third colliding molecule takes away a little of the energy, as in Figure 2-21, the iodine atoms will not quite be able to separate. Later, perhaps another collision will drain off more energy, and the vibrational movement of the iodine atoms will be cooled still further. After a time and after many collisions, the iodine atoms will find themselves at the bottom of the valley, at the equilibrium distance. Their onetime energy will have been transferred into the random energy (heat) of the collisional partners that made it all possible.

This is not just a fiction about friction. Experimentally, it is found that iodine atoms react very slowly, even though the formation of I_2 releases 36 kcal/mole as the reaction proceeds. They react slowly, that is, if the pressure is low. But as the number of collisions per second increases, the rate at which I_2 is formed increases. Molecular friction drains off the ordered energy and converts it into heat.

(c) SO WHAT ABOUT SPONTANEOUS CHANGES?

So why do our ball bearing and our two iodine atoms end up at the bottom of the hill? Because friction tends to convert ordered motion into disordered motion. That seems to be involved in

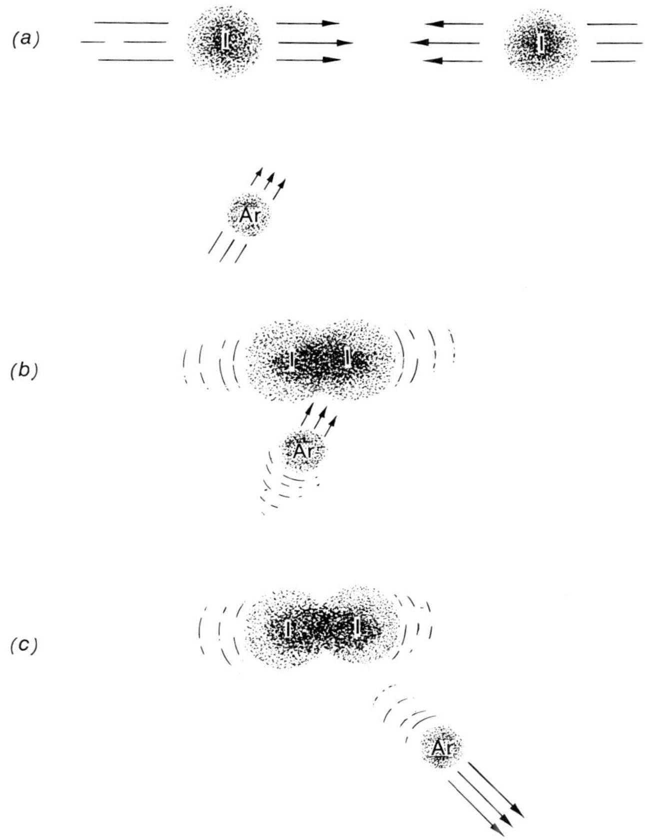

Figure 2-21 Collisional friction helps iodine atoms react: An argon atom will do.

spontaneous change. Yet we still haven't grasped the essence of the matter, for some changes occur spontaneously, absorbing energy as they take place. Can friction account for that? No, indeed.

So we conclude that the generation of disordered energy (heat) from ordered energy (either potential energy or kinetic energy) is somehow involved in spontaneous change, but not simply through the tendency of things to roll downhill. How do the molecules know what to do? Maybe we should choose some very simple spontaneous process and watch these intelligent molecules, so we can learn. The process we'll begin with is the expansion of a gas into a vacuum. This simple process gives insight into the mass psychology and combined intellect of a mole of molecules, as they inexorably head for equilibrium. We shall find that molecules *do* roll downhill, but not on an energy landscape.

three
randomness and chemical change

In Chapter One we stressed the importance of the equilibrium state and, particularly, its composition. This is given by the equilibrium constant which measures the extent of reaction. Chapter Two dealt with the energy changes that accompany chemical changes, but did not come to grips with the question of when a reaction is spontaneous. Now we will probe into this problem and discover the various factors that influence the position of the equilibrium state. We already know that a knowledge of the energy change ΔE or ΔH is not a sufficient condition for predicting the direction of chemical change.

We will begin by looking at a very simple process: the expansion of a gas. The energy changes associated with such a process are extremely small (zero for a perfect gas), so we will gain some insight into the factors other than energy which affect chemical change.

3-1 Expansion of gas

Figure 3-1 shows two identical pieces of apparatus; each consists of a 1-liter bulb connected through a closed stopcock to a 3-liter bulb. In apparatus A, the 1-liter bulb A_1 contains Cl_2 at a pressure of 24 torr. Bulb A_2 is empty. In apparatus B, the 1-liter bulb B_1 has been filled with Cl_2 to a pressure of 3 torr, and the right-hand bulb B_2 has been filled to 7 torr.

Now, consider the question—what will happen in each of these systems if the stopcocks are opened? The situation is such a familiar one that none of us has the slightest difficulty in confidently predicting the result. In A, the Cl_2 gas will move from the smaller bulb A_1 into the larger bulb A_2 until the total pressure is uniformly 6 torr. In apparatus B, gas will move in the opposite direction, from the larger bulb to the smaller bulb, again until the pressure in the system is equalized at 6 torr. Everyone knows that a gas expands into a lower-pressure region until pressure equilibrium is reached. Despite our confidence, we try the experiment. The stopcocks are opened and the predicted changes occur, as shown in Figure 3-2. So our question was an easy one. Only very little thought told us what behavior to expect.

Now let us visualize this same "experiment" on

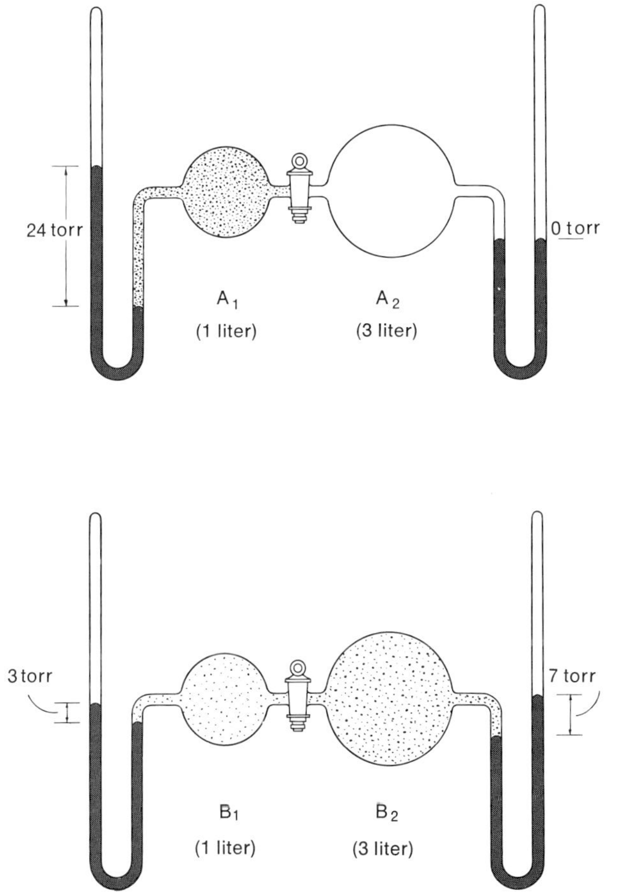

Figure 3-1 *What will happen when the stopcocks are opened?*

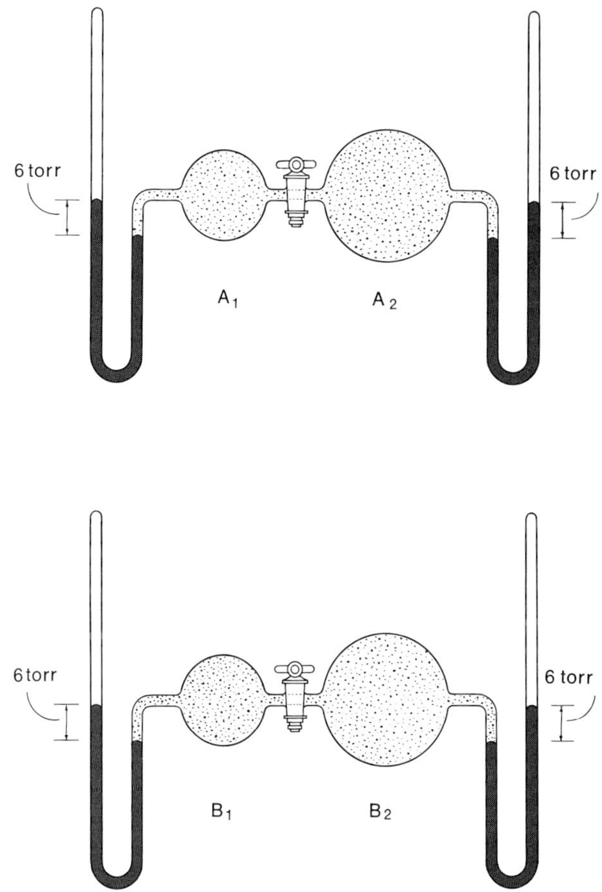

Figure 3-2 *The gas pressure is uniform.*

a microscopic level. Picture molecules of Cl_2 dashing about in the left-hand bulb of apparatus A and in the left-hand bulb of apparatus B. The stopcocks are now opened. In a short interval of time, $\frac{3}{4}$ of the molecules in A_1 somehow decide to move into A_2, whereas, during the same period, molecules from B_2 move in the opposite direction into B_1. How did the molecules in A_1 *know* they should move into A_2? Were they, as we were, thinking back to past experience to decide? Not likely! Many scientists doubt that molecules think at all—yet molecules always do the right thing.

This poses a harder question. What governs the behavior of a mole of gas? How can we explain its reproducible performance in a given set of circumstances despite our general disrespect for the molecular mentality? *What are the factors that determine the direction of a spontaneous change?*

(a) AN EXAMPLE—RANDOMNESS AT WORK

Figure 3-3 shows a gambling game. The board has 25 depressions, numbered one to 25. A player bets one dollar and selects three numbers, say 3, 7, and 12. The operator then throws three ping-pong balls into the box. If the balls settle into the correct depressions, those numbered 3, 7, and 12, the house pays the player \$2000. This enormous reward is, however, much less enticing than that in game 2 pictured in Figure 3-4. Here there are 100 depressions. This time the player who bets one dollar and

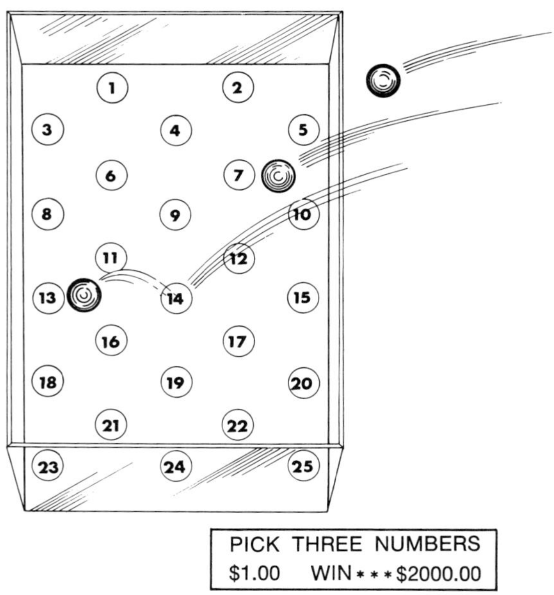

Figure 3-3 Randomness game number one.

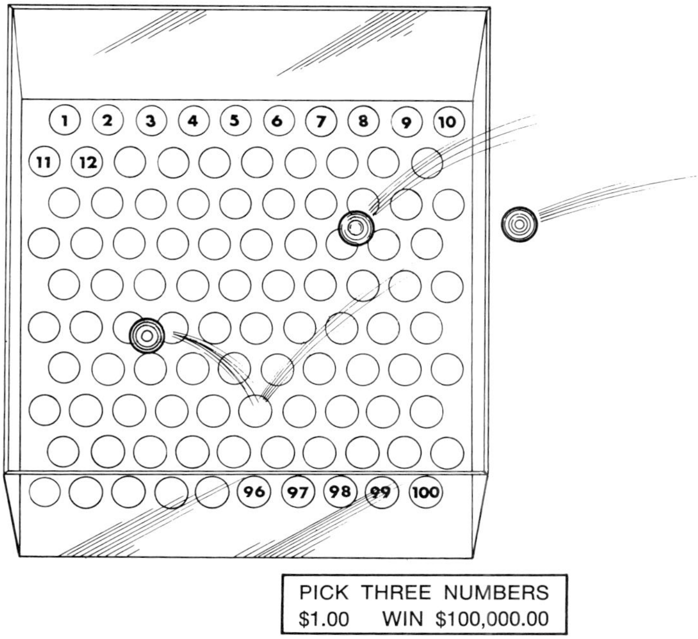

Figure 3-4 Randomness game number two.

picks the right three numbers wins $100,000. This is the way to get rich!

Not really! The man who runs the game is the only person who can count on getting rich. When the evening is over and thousands of bets have been made, the three ping-pong balls will have won for the house 24 percent of all the money bet on game 1, shown in Figure 3-3. At the other table, with its enormous payoff, the three ping-pong balls will have won 40 percent of the players' money for the house!

Apparently, these ping-pong balls are intelligent, and if so, the operator is using the smarter balls in game 2. But you know and the operator knows that ping-pong balls cannot be trained; nor are they loaded, or magnetized. The operator merely understands positional randomness. With this knowledge, he has carefully calculated the probability of any three particular numbers being obtained, to determine the payoff in each game and to guarantee his own success. Quite contrary to our implication of wile on the part of the ping-pong balls, the operator requires that they be stupid—they must be absolutely random in their choice of position, lest the players learn their preference. All the house wants is for the ping-pong balls to choose their positions at random and then the results of the game are perfectly predictable. Over time, the house always wins!

(b) BACK TO MOLECULES—ARE THEY STUPID TOO?

Could there be a connection between the experiments of Figure 3-1 and the performance of the ping-pong balls? Could molecules be as unenlightened as ping-pong balls, picking their positions at any given instant completely at random and without the slightest preference? As we ask this question, we are playing another game, the game of science. We have observed the behavior of a part of our environment, the expansion of a gas into a vacuum, and wondered about it. We then looked around for something else in our experience that is better understood, and came up with the gambler's lucrative game. With this possible analogue, we have posed the scientist's question: "Does the ping-pong ball game furnish a *model* with which we can understand the expansion of a gas into a vacuum?"

What is the next step in this game of science? Having proposed a model, we must explore its implications—*we must test the model.* Without such tests, science would be only a meandering of the mind, as it was in the time of Aristotle—as it was until the time of Galileo who not only observed and pondered, but also tested his ideas.

Suppose that at any instant of time, each molecule takes a position that is a completely random and impartial choice of all possible positions accessible to it. What would this model predict for the experiments in Figure 3-1? This question can be answered only in probability terms. If we had but one molecule, it would be found in the 1-liter bulb one quarter of the time and in the 3-liter bulb three quarters of the time. If we had four molecules to worry about, the probability that all four would be in the 1-liter bulb is 1/256, while the probability that two would be in each bulb is 54/256. An arrangement of special interest to us is that in which one molecule is found in A_1 and three in A_2, which can be calculated (with our dumb-molecule model) to have a probability of 108/256. *This is the most probable arrangement.* If this type of calculation is made for 16 molecules, the most probable arrangement proves to have 4 molecules in A_1 and 12 in A_2 (again in the ratio 1:3). On a comparative basis, the arrangement of all 16 molecules being in A_1 is only one in 10^9 as likely. The arrangement of 2 molecules in A_1 and 14 in A_2 is only 54/91 as likely. If we take more and more molecules, the arrangement in which A_1 contains one third as many molecules as A_2 is *always* the most probable of *any* possible arrangement. Relative to this probability, all other arrangements become less and less probable, including the arrangements in which all the molecules are in A_1, or $\frac{1}{8}$ in B_1 and $\frac{7}{8}$ in B_2, the distributions shown in Figure 3-1. As the molecular population increases, the probability distribution becomes more and more narrow, always peaked at the distribution that has the molecules distributed in the ratio of the volumes of the two bulbs: 1:3, as shown in Figure 3-5.

Now we can see where this model will lead us. As we deal with

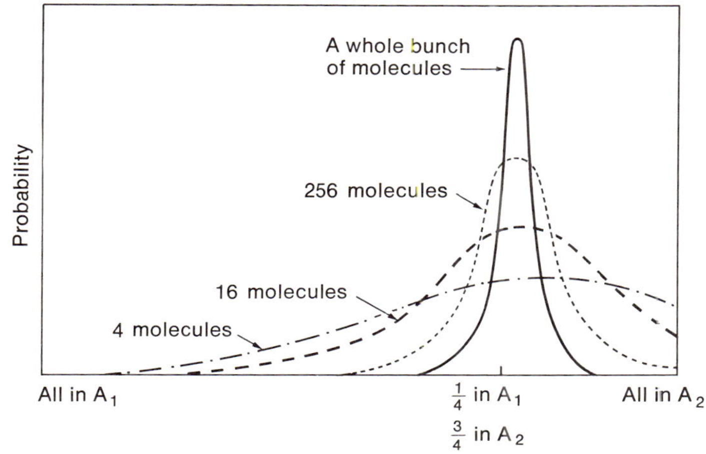

Figure 3-5 How the probability of molecular distribution depends on the number of molecules (random position model).

larger and larger numbers of molecules (approaching Avogadro's number), *the probability* will close in on a situation in which *the molecules are distributed between the two bulbs in proportion to the bulb volumes.* This is the distribution that makes the number of molecules per milliliter the same in the two bulbs. This is exactly the condition needed to give equal pressures.

It is also profitable to examine the reverse of this process. Why don't the gas molecules of Figure 3-2 all rush back to the initial arrangement in Figure 3-1, at least occasionally? Once again we must answer in probability terms. For this event to occur, every single molecule must move in the same direction—through the stopcock into the left-hand bulb. It is clear that a situation in which several billion randomly moving molecules all pack up their bags and set off in the same direction is highly improbable. So, given a choice, molecules will find the state of greatest positional randomness and *stay there.*

So we can understand the changes which take place from the initial conditions in Figure 3-1 to the final ones in Figure 3-2 to be spontaneous changes towards the most probable positional arrangement. Our model, based upon positional randomness for each molecule, predicts the direction of spontaneous change for the expansion of a perfect gas.

3-2 An exothermic chemical reaction

Figure 3-6 shows a container which is separated into two halves by a thin membrane. The lower half contains hydrogen at a pressure of 10 torr, whereas above the membrane there is fluorine at a pressure of 10 torr. At the top there is a small weight held by

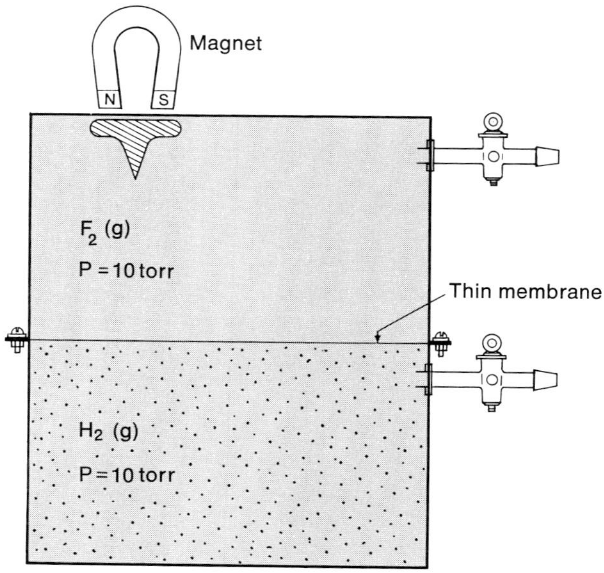

Figure 3-6 What happens when hydrogen and fluorine mix?

an external magnet. When the magnet is removed, the weight drops, pierces the membrane, and allows the gases to mix. Will they spontaneously react to form hydrogen fluoride?

$$H_2(g) + F_2(g) \overset{?}{\rightarrow} 2HF(g) \qquad (3\text{-}1)$$

Once again, many can answer this question without hesitation. Fluorine has a wide reputation as a fearsomely reactive chemical (iron ignites and burns when in contact with liquid fluorine), and we found in Chapter Two that its reaction with hydrogen liberates a large amount of heat (129 kcal per mole of fluorine reacted). Again, we have asked an easy question.

Once again, the more penetrating question is to query why the atoms spontaneously decide to rearrange to form the more stable molecules of hydrogen fluoride. Having decided that molecules are probably *non compos mentis,* it seems likely that atoms are no more intelligent. How can we understand their decisive behavior?

Let's take a closer look at this reaction mixture. As a mole of F_2 reacts with a mole of H_2, a large amount of energy is liberated as heat: the atoms take up more favorable positions, energy-wise, as they form HF molecules. We can think of the energy associated with a given molecular arrangement as potential energy (energy of position). The potential energy of two moles of HF molecules is lower than that of the same number of atoms arranged as a

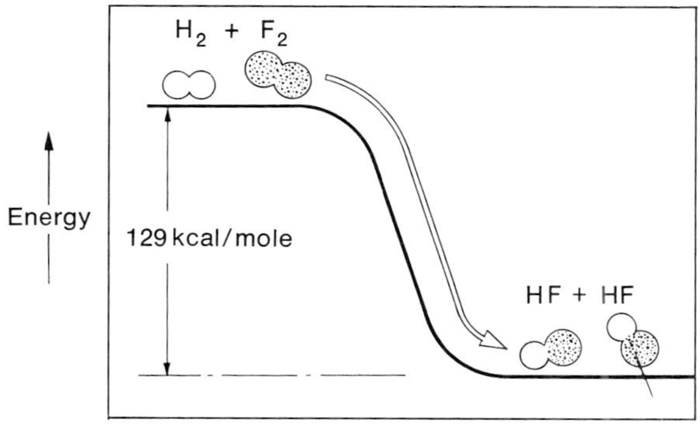

Figure 3-7 Potential energy change during the reaction
$H_2(g) + F_2(g) = 2HF(g)$
On an energy landscape the reactants roll downhill to form products.

mole of H_2 molecules and a mole of F_2 molecules. As HF is formed, the atoms are "rolling downhill." The potential energy change results in the appearance of thermal energy (Fig. 3-7).

Yet again we encounter our question: "*Why* do molecules tend to roll downhill?" Why would the atoms end up in the energetically favorable positions? There is no energy "gain"—after all, energy is conserved. Someone merely has to figure out how to divide up the energy released among the random translational motions of the product molecules. Think of the confusion as this energy is somehow partitioned among the many molecular claimants.

In fact, here is the seed of an idea. Perhaps the model of random behavior that explained the spontaneous gas expansion will also furnish a basis for explaining why molecules "roll downhill."

(a) A MODEL—ENERGY PACKETS RANDOMLY DISTRIBUTED

section 3-2 exothermic chemical reaction

75

If one mole of F_2 and one mole of H_2 react according to (3-1), we find 129 kcal of energy released. Two moles of HF are formed, and these product molecules must divide up the 129 kcal into translational, rotational, and vibrational energy. One way this could be done would be to give every molecule an equal share—$129/2N_0$ (N_0 is Avogadro's number). In fact, the energy was almost divided in this way before the reaction, when it was in the form of potential energy. Every reactant molecule (H_2 or F_2) possessed the same potential energy as any other like reactant molecule. From a probability point of view, there would be little to choose between reactants and products if every product molecule possessed the same kinetic energy as every other.

There is another way in which energy of motion can be apportioned, however. Suppose the 129 kcal are divided into small increments, and that these energy packets are then distributed among the molecules at random, one at a time. This would be like collecting three coins each from 100 children, and then redistributing these 300 coins by throwing them, one at a time, into the crowd. There is only a slight chance that each child would end up with three coins. It is far more likely that some children would catch several coins, and some unlucky ones would receive none, only one, or only two coins. From an economic point of view, this might be inequitable, but from a statistical slant, such a coin distribution is more probable. If the 129 kcal were similarly distributed in small increments to the molecules, paying no attention at all to a given molecule's previous energy "wealth" or "poverty," it is most probable that an "inequitable" distribution would occur. Some molecules would have lots of rotational, translational, and vibrational energy, many would have energies near the average, and some would have very little or none at all. According to this model, *motional energy is distributed in the most probable way.* Considering translational motion first, this model predicts a range of velocities for the molecules in a gas. This prediction can immediately be compared with the experimental evidence, which has been known for many decades. Molecular velocities of gas molecules *are* distributed over a range—they exhibit the famous "Boltzmann distribution" of velocities shown in Figure 3-8. At least in a qualitative way, we can see that a tendency towards motional randomness may be connected with spontaneous changes.

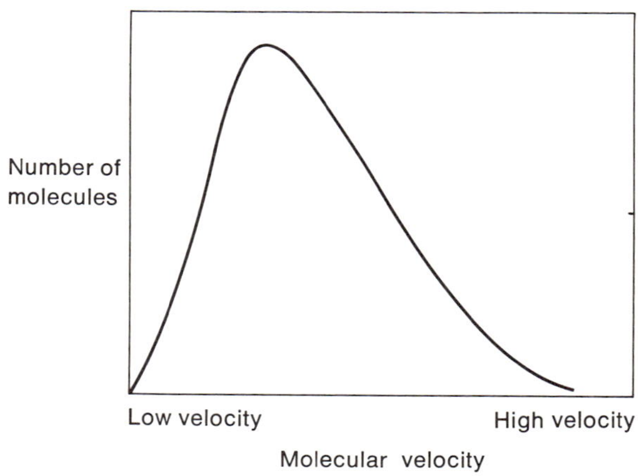

Figure 3-8 Experimentally determined velocity distribution of gas molecules.

A more quantitative (though still approximate) check on this model can be developed. Let us consider the distribution of "packets" of translational energy among molecules.

(b) A QUANTITATIVE EXAMPLE: VELOCITY DISTRIBUTION IN A GAS

The kinetic theory of gases accounts for the properties of a gas in terms of the directions and velocities of the molecular motions. According to this theory, a mole of a perfect gas, at temperature T (°K), has a translational energy of (3/2)RT.

Imagine that the gas is contained in a cubic box. The gas pressure is the same on every face, so that, on the average, the translational energy is (1/2)RT in each of the three (x, y, z) directions. Since there are N_0 molecules in a mole, each molecule has, on the average, kinetic energy equal to

$$\frac{1}{N_0}\left(\tfrac{1}{2}RT\right) \text{ erg} \tag{3-2}$$

associated with motion in the x direction, and the same amount, on the average, in each of the other two directions. Since we are often interested in the energy of a single molecule, it is convenient to define a new constant

$$k = \frac{R}{N_0} \tag{3-3}$$

which relates molecular energy directly to the temperature:

$$E_{\text{av.}} = \tfrac{3}{2}kT \text{ erg per molecule} \tag{3-4}$$

or

$$E^x_{\text{av.}} = E^y_{\text{av.}} = E^z_{\text{av.}} = \tfrac{1}{2}kT \text{ erg per molecule} \tag{3-5}$$

With this in mind, let's consider a simple system of four molecules which would have, at temperature T, a total translational energy of 4(3/2)kT erg. This amount averages out correctly to (3/2)kT erg per molecule. Suppose, however, that we divide the (12/2)kT erg into many packets, and then proceed to distribute these packets at random, "throwing them, one at a time, into the crowd." One can sense that the packets should be extremely small compared with the average energy. However, we'll make life simple by taking only a few packets, so we can keep track of our energy lottery. We'll take each packet to be $\frac{1}{2}kT$, so that only 12 packets must be accounted for. Even with these unrealistically large packets, there are many ways in which the four molecules might distribute the energy, but few enough for us to write down some of them in Table 3-1. We see that there are only four ways in which one molecule could get all 12 packets, but there are 12 ways in which one molecule could get 11 packets with some other molecule receiving the remaining one packet. The latter is a more probable distribution of the 12 packets. Still more prob-

Table 3-1 Some Ways in which Twelve Energy Packets can be Distributed among Four Molecules

Molecules			
A	**B**	**C**	**D**
12	0	0	0
0	12	0	0
0	0	12	0
0	0	0	12
total = 4			
11	1	0	0
11	0	1	0
11	0	0	1
Plus 9 more in which B, C, or D has 11			
total = 12			
10	1	1	0
10	1	0	1
10	0	1	1
10	2	0	0
10	0	2	0
10	0	0	2
Plus 18 more in which B, C, or D has 10			
total = 24			

able is the distribution which gives one lucky molecule 10 packets, with two packets somehow doled out to the other three fellows—and so it continues. We can't say which distribution will occur; any one is as probable as any other, as far as our model goes.

Despite our lack of progress, let's persist and consider, now, Avogadro's number of molecules. This would really require too long a table to reproduce here, or, in fact, on all the paper in existence. However, we can approximate the situation by taking the N_0 molecules four at a time and, to keep track of energy, require that each group of four molecules divides up its "share" of $4(3/2)kT$ ergs of translational energy according to our model. If every set of four molecules averages $(3/2)kT$ erg per molecule, then surely the entire crowd of N_0 molecules will also average $(3/2)kT$ erg per molecule, or $(3/2)RT$ erg/mole.

Now statistics come to our rescue. If we have $\frac{1}{4} \cdot 6 \cdot 10^{23}$ quartets of molecules, we can be confident that every one of the arrangements shown in Table 3-1 will be represented many, many times. Since each single arrangement is as probable as any other arrangement, we can take the superposition of all of these arrangements to be an approximate prediction of the properties of the mole of gas. It will be approximate because we considered the molecules four at a time as a labor-saving device.

Now we can make a prediction: our model predicts the distribution of molecular energies. If we reach into the gas and buttonhole a molecule at random, what is the probability that we will get one with no energy packets? Or with one energy packet? Or with any number from zero to twelve? The answer is obtained merely by counting the number of times each possible energy is found in a complete version of Table 3-1, since every arrangement will occur as often as any other. The result of this count is shown in Table 3-2. The most likely energy is predicted to be zero, and the least likely is 12—many arrangements give one or two molecules zero energy, whereas only the first four give one molecule as many as 12 packets of energy. We can perceive the results more easily in the graphical presentation in Figure 3-9. We see that the energy distribution resembles an exponential one.* In fact, if we consider six, eight, and ten molecules in the way we treated four in Table 3-1, the energy distributions approach more and more closely to an exponential curve.

Table 3-2 Distribution of Molecular Energies: A Mole of Gas Considered Four Molecules at a Time

Number of Energy "Packets" i	Number of Occurrences n_i	Fractional Number of Occurrences $n_i/\Sigma n_i$
0	364	0.200
1	312	0.171
2	264	0.145
3	220	0.121
4	180	0.099
5	144	0.079
6	112	0.062
7	84	0.046
8	60	0.032
9	40	0.022
10	24	0.013
11	12	0.006
12	4	0.002
	$\Sigma n_i = 1820$	

Now we can return to the experimental data shown in Figure 3-8, but expressed in terms of molecular velocities. We see that there is no immediate resemblance between the predictions of our model and the experimental facts. The "theory" developed from our model does not agree with the facts of nature.

At this juncture in the development of a theory, a scientist has two options. He can cast about for an entirely different model ("theory"), or, if he has sufficient confidence in the premises of the inadequate model, he can investigate minor changes in the

*Exponential curves are discussed more fully in Appendix A.

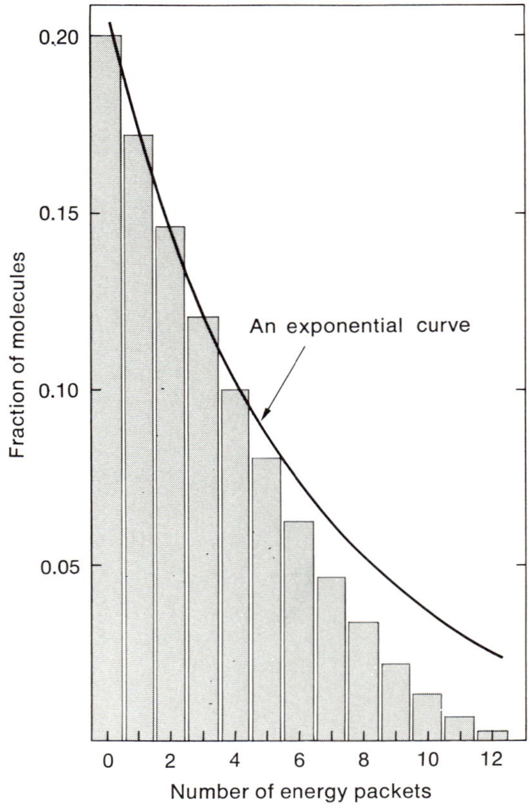

Figure 3-9 Molecular energy distribution corresponding to a random allotment of energy packets.

model that still retain the desired premises. In our problem, a useful theory was devised by the latter approach.

(c) A PATCH ON OUR MODEL OF RANDOM ENERGY DISTRIBUTION

The random distribution of small packets of energy leads to an exponential distribution. Could there be another factor at work that would convert our exponential curve of Figure 3-9 into the peaked and asymmetric experimental curve of Figure 3-8? We are looking for a factor that reduces the probability of low molecular energies and we would like it to be implied by the packet view of energy distribution.

The answer proves to be surprisingly simple—though conceptually quite significant. As a molecule gains "packets" of energy in our energy lottery, its velocity goes up in discrete jumps. However, the velocity of a molecule defines its momentum, since momentum equals mass times velocity. Hence, momentum must also come in "packets." To see how this affects our model, we must recall that for any given velocity, a molecule can move in a variety of directions. The variety of directions, which determines the probability of a given momentum, will be limited by the "packet" view we are assuming. This can be seen pictorially with a building-block model. In two dimensions, the accumulation of translational momentum might occur as in Figure 3-10. As we add the first momentum increment, we can "pack" it in any of several positions—eight are shown in the figure. Adding another momentum increment increases the number of positions into which the second increment can be packed—there are more "directions" now—16 are shown. With three increments, the number of "directions" becomes 24. In two dimensions, the number of directions is proportional to the number of momentum increments.

Extending this same building-block model into three dimensions adds no new ideas. However, as shown in Figure 3-11, the number of directions goes up more rapidly as momentum increments are added. In fact, now *the number of velocity directions depends upon the square of the number of momentum increments.* Our exponential energy lottery model must be modified by a factor of v^2 to permit it to fit into the "momentum space" pictured in Figure 3-11. The probability $n(v)$ that a particle at velocity v will have a certain energy is a product; an exponential factor fixed by the random distribution of energy packets, but weighted by a v^2 multiplier that takes into account the available "momentum space." Hence,

$$n(v) = cv^2e^{-[(1/2)mv^2/kT]} \qquad (3\text{-}6)$$

This equation proves to be in exact agreement with the experimental velocity distribution shown in Figure 3-8. The proportionality constant c will not be discussed,* but it does not alter the conceptual content of our now-successful model.

(d) SUMMARY—RANDOMNESS IN A MONATOMIC GAS

Now our model fits the properties of a monatomic gas. This is a particularly simple system because there are only positional and translational movements to worry about (no vibration or rotation). For such a simple system, energy distribution among the possible translational movements occurs in a random fashion, the ultimate distribution among a huge number of molecules being the most

*Statistical mechanical arguments show that $c = 4\pi N(m/2\pi kT)^{3/2}$. This constant actually relates to the magnitude of the translational energy packets, and they prove to be small, indeed. Instead of $\frac{1}{2}kT$-sized packets, they must be 10^{-18} times smaller. Nevertheless, the model shows that translational energy must be quantized as required by quantum theory.

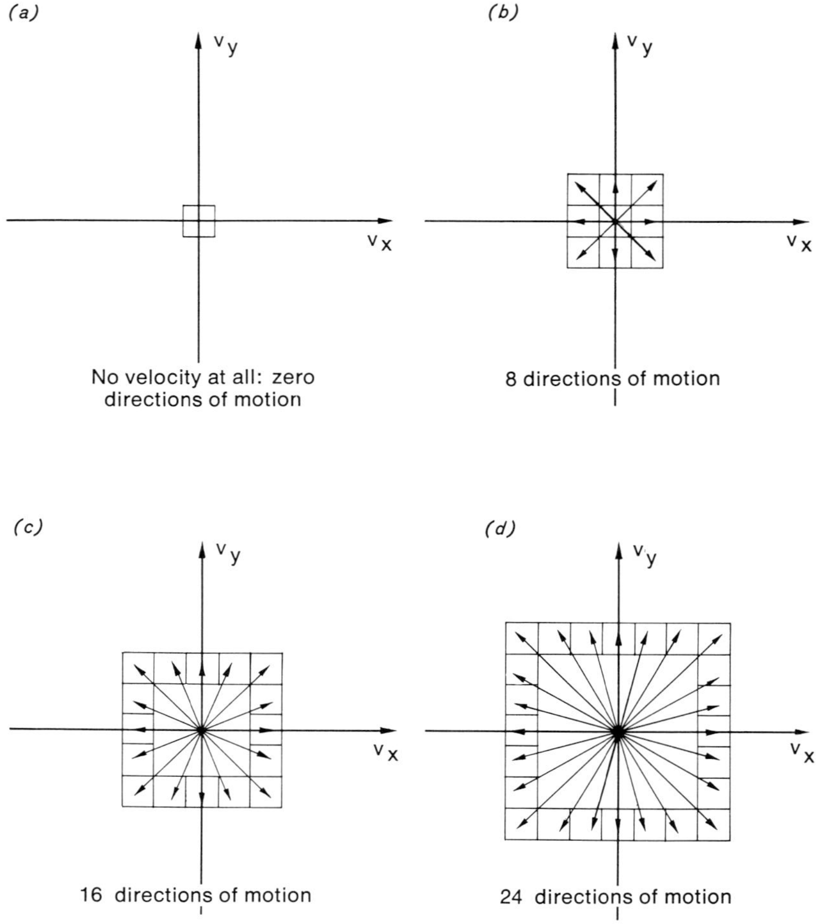

Figure 3-10 (a) *A molecule with zero energy;* (b) *a molecule with one increment of momentum;* (c) *a molecule with two increments of momentum;* (d) *a molecule with three increments of momentum.*

probable one. Again, as in the case of positional randomness, we find that a spontaneous change will take the direction that increases the randomness of energy distribution among the available repositories for energy—that is, among the available degrees of freedom.

Thus we have found an interesting parallel between positional and motional randomness. We could explain the spontaneous expansion of a gas if we assumed that all the positions in our three-dimensional positional space are equally probable and that the system tends to change to the most random (hence the most probable) positional arrangement. In the case of energy release, as in an exothermic chemical reaction, we found that the

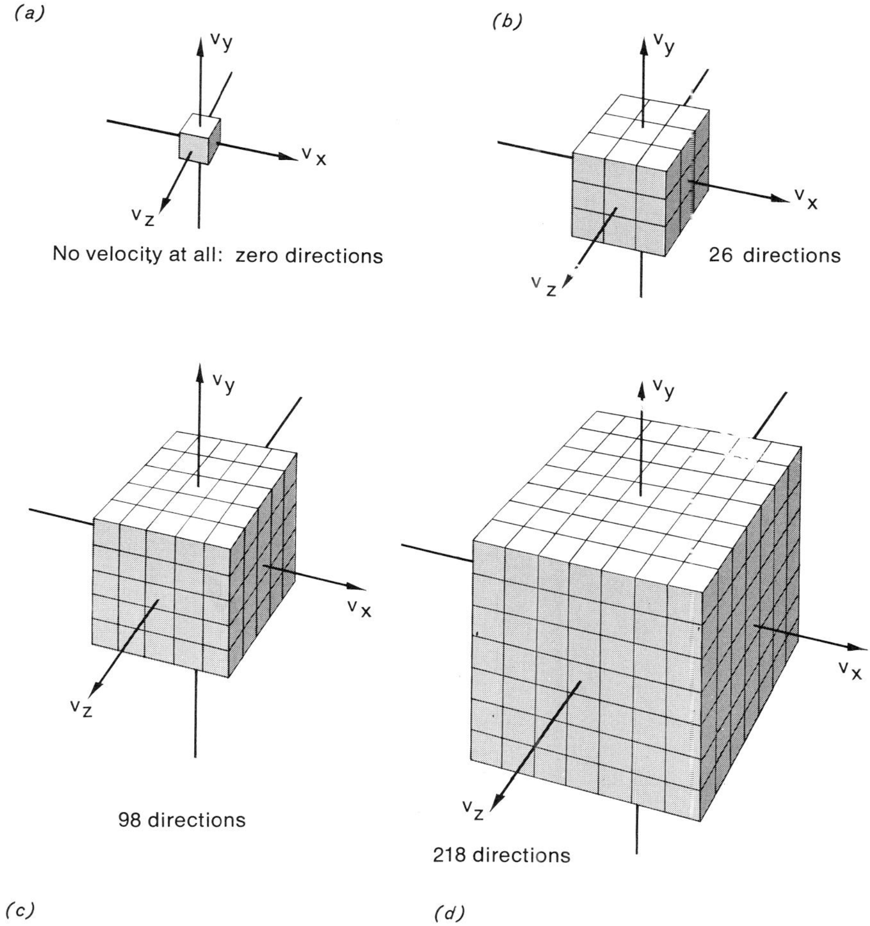

Figure 3-11 (a) *A molecule with zero momentum;* (b) *a molecule with one increment of momentum;* (c) *a molecule with two increments of momentum;* (d) *a molecule with three increments of momentum.*

distribution of this energy could be explained in a similar way. Here we assumed that all "velocity positions" in a "three-dimensional velocity space" (as shown in Fig. 3-11) are equally probable, and that the system tends to change to the most random (hence, the most probable) velocity arrangement. This most probable velocity arrangement also means the most probable energy arrangement.

This parallelism is usually expressed in terms of momentum rather than velocity. The momentum p of a particle is just its mass times its velocity:

$$p = mv \tag{3-7}$$

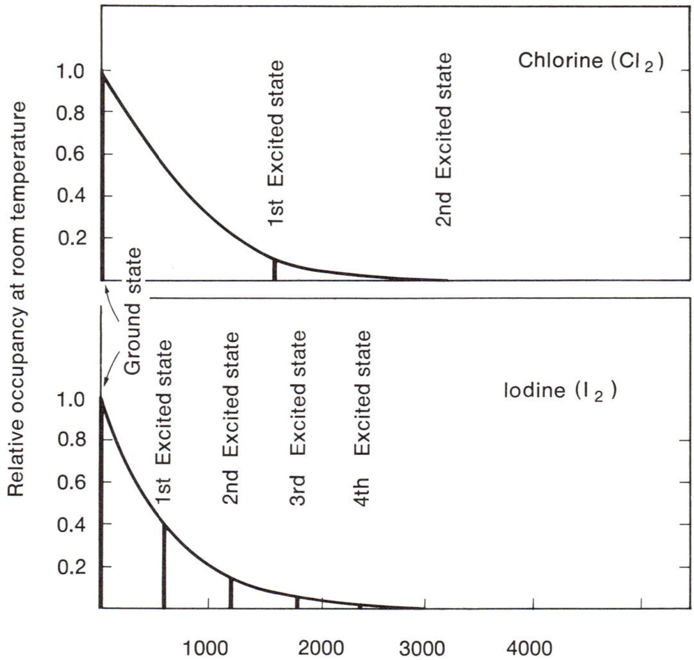

Figure 3-12 Relative occupancy of Cl_2 and I_2 vibrational states at 300°K.

Hence, our randomness picture for a monatomic gas can be expressed in a single expression if we consider a "six-dimensional space" for each atom whose six "Cartesian coordinates" are three positional coordinates *x,y,z,* and three momentum coordinates p_x, p_y, p_z. In this space, called "phase space," there are six coordinates for each molecule. Now we can make a single statement about randomness: *a system tends to change spontaneously towards the most random (the most probable) arrangement in phase space.*

(e) RANDOMNESS IN A POLYATOMIC GAS

These same ideas of randomness apply equally well to polyatomic molecules, with their additional vibrational and rotational energy states. These states are quantized and, in the case of vibrational states, they are quite widely spaced. Needless to say, a particle in a high-energy state has a relatively small chance of being excited, for the same reason as finding all twelve energy packets cornered by one molecule was unlikely, in our energy lottery. Concentration of the energy is not conducive to randomness. In fact, the exponential weighting factor again determines the occupancy of these states. There is no analogue to the directional

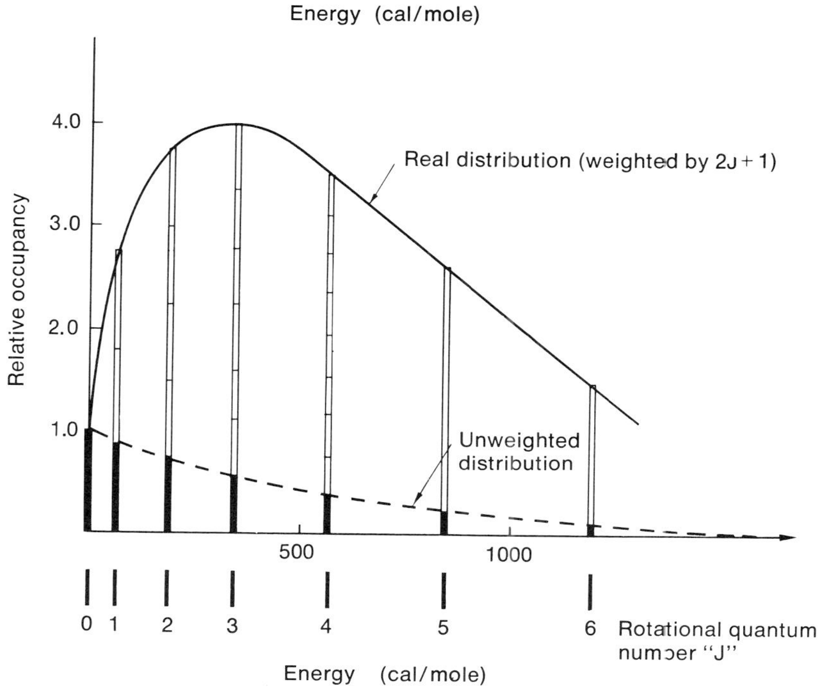

Figure 3-13 Relative occupancy of HCl rotational states at 300°K.

v^2 factor for vibrational states, so the occupancy merely follows the exponential factor on an intermittent basis. Figure 3-12 shows the relative occupancy at room temperature for the excited vibrational states of iodine, with a rather small energy spacing, and of chlorine, with a larger spacing. For iodine, the first excited state has about one third the occupancy of the ground (unexcited) state. In contrast, the vibrations of chlorine are barely excited at room temperature. This latter situation is more normal for simple molecules, but polyatomic molecules generally have a few vibrational movements with frequencies as low as those of iodine. Nevertheless, it is a reasonable approximation to assume that, at room temperature, the molecular vibrational states make little contribution to energy randomness.

The situation is quite different for rotation. First, the levels are much closer together, and secondly, there is a weighting factor analogous to the v^2 term (equation (3-6)). Figure 3-13 shows the occupancy, at room temperature, of the HCl rotational levels. Each level is identified by an integer quantum number J, and each rotation can be oriented in $(2J + 1)$ directions (the analogue to v^2). Hence, on a scale where the occupancy of $J = 0$ is called unity, the $J = 3$ state has an occupancy of four despite the exponential factor. Even states above $J = 6$ are more heavily populated than $J = 0$. Hence, the rotational states contribute very significantly to the energy randomness.*

The most obvious experimental evidence for the contribution of rotation and vibration to energy randomness is manifested in the heat capacity of the molecules. Table 3-3 gives the heat capacities of some gaseous substances and the breakdown into the various contributions. Plainly the rotational states play a heavy role in the energy lottery—as important as the translations—and for the heavier diatomic and for polyatomic molecules (which have low frequencies), the vibrational excitation can also be an important term.

Table 3-3 Heat Capacities at Constant Volume: Contributions by Rotation and Vibration (cal/mole degK)

Molecule	C_v(300°K)	=	Translational Part	+	Rotational Part	+	Vibrational Part
He	2.98	=	2.98	+	none	+	none
Ar	2.98	=	2.98	+	none	+	none
HCl	4.97	=	2.98	+	1.99	+	0.0004
Cl_2	6.12	=	2.98	+	1.99	+	1.15
I_2	6.82	=	2.98	+	1.99	+	1.85
CH_4	6.55	=	2.98	+	2.98	+	0.59
CCl_4	17.97	=	2.98	+	2.98	+	12.01
C_2H_6	10.60	=	2.98	+	2.98	+	4.64

In summary, when polyatomic molecules participate in a chemical change, any energy released must be distributed among the products, paying attention to all degrees of freedom available: translation, rotation, and vibration.† This distribution occurs with all possible energy states competing on a lottery basis, in accordance with the exponential energy–temperature dependence and with weighting factors appropriate to the degree of freedom. These considerations lead to the most random energy distribution—the one nature is searching for.

(f) ENERGY AND POSITION IN COMPETITION

Here are three exothermic chemical reactions:

$$NH_2 + NH_2 \rightarrow N_2H_4 \quad \text{heat released} = 56 \text{ kcal} \qquad (3\text{-}8)$$
$$NF_2 + NF_2 \rightarrow N_2F_4 \quad \text{heat released} = 20 \text{ kcal} \qquad (3\text{-}9)$$
$$NO_2 + NO_2 \rightarrow N_2O_4 \quad \text{heat released} = 14 \text{ kcal} \qquad (3\text{-}10)$$

In each reaction, the reactants have higher potential energy than the products. Each reaction is "rolling downhill" as far

*The vibrational and rotational results come directly from the application of quantum theory to the motions of atoms in molecules.

†We have omitted reference to electronic excitation but only a few molecules have such low energy excited electronic states that they, too, must be considered.

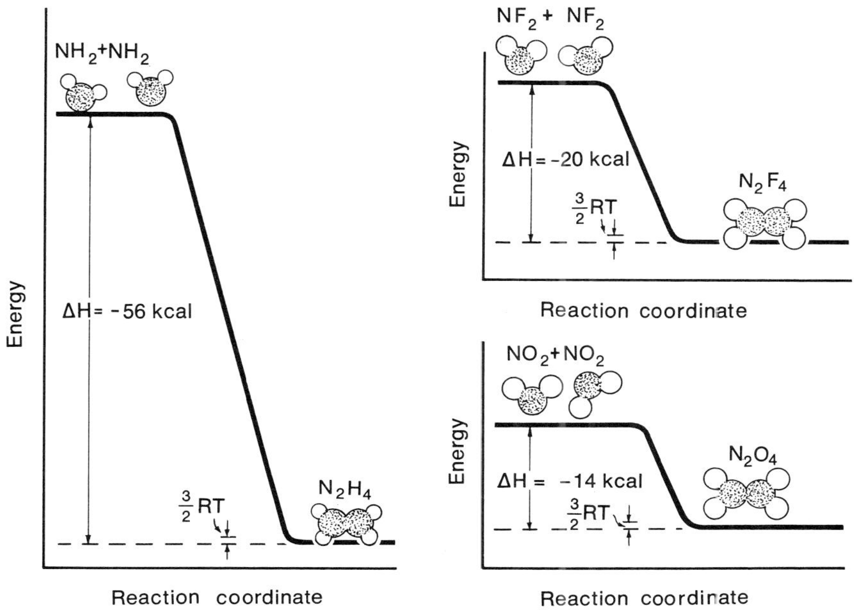

Figure 3-14 Three Similar Exothermic Reactions.

as energy is concerned (see Fig. 3-14), but now we know that randomness is what matters. Energy randomness is only one of the factors at work—positional randomness must also be considered.

We'll consider the energy randomness first. In each of these exothermic reactions, there is an increase in randomness connected with taking an amount of potential energy from each reactant and distributing it at random among translational, rotational, and vibrational degrees of freedom (i.e., releasing it as heat). Insofar as this effect is concerned, we are led to expect that all three reactions will occur.

Let's try the experiment. We place 0.40 moles of NH_2 into a 24.5-liter bulb (never mind how we do it) immersed in a constant temperature bath at 298°K. The pressure is measured to be 0.20 atmospheres. This is exactly the pressure expected if all of the 0.40 moles of NH_2 molecules react according to (3-8) to form 0.20 moles of N_2H_4. Our expectation is realized. The very large increase in randomness associated with divvying up 56 kcal/mole among the translational, rotational, and vibrational degrees of freedom in the constant temperature bath causes the reaction to proceed with great, chaotic enthusiasm.

Now, let's do the same experiment with 0.40 moles of NF_2 molecules in a second 24.5-liter bulb and 0.40 moles of NO_2 in a third bulb of the same volume. In the second bulb, we measure a total pressure of 0.202 atmospheres, slightly above the 0.200 atmosphere pressure obtained with N_2H_4. Apparently the reaction is not quite complete! In the third bulb, the effect is even more significant. Here, the pressure is found to be 0.26 atmospheres, the pressure appropriate to the formation of only about 65 percent of the possible number of N_2O_4 molecules! Apparently, only about two-thirds of the NO_2 molecules decided to "roll downhill" and the other one-third refused to join them!

Our first thought might be to place the blame on the NO_2—perhaps we have the wrong reaction. Not so, however. Spectroscopic studies (the absorption of light by the gas mixture) clearly corroborate our interpretation of the pressure. The nitrogen dioxide *does* contain about equal numbers of the NO_2 and of N_2O_4 molecules; it reached two-thirds completion and stopped rolling downhill! Hence, positional randomness must be at work.

To verify this interpretation, let's repeat all three experiments at 400°K. At this higher temperature, the NH_2 molecules again react almost completely to give hydrazine, N_2H_4, and the released energy is divided up into translational, rotational and vibrational energy packets. In the NF_2 bulb, however, we find that at 400°K only 60 percent of the N_2F_4 molecules have been formed! Now 40 percent of the NF_2 molecules refuse to roll downhill! In the NO_2 bulb, the pressure is very close to 0.40 atmospheres—almost no reaction occurs at all!

We see from these results that we cannot ignore positional randomness, and that its importance increases as temperature goes up. Just as when a bachelor decides to take the big step, when two NO_2 molecules join together in the holy state of chemical bondedness, the loss of freedom of movement is considerable. Whereas before bonding the two molecules could behave completely independently, both from a position and an energy point of view, now, as an N_2O_4 molecule, they must share the same energy bank account, live closely together in the same geographical location, go to the same molecular parties, and so on.

It is intrinsically more random to have many molecules rather than a few (for a given of atoms). It is also intrinsically more random to have several kinds of molecules rather than one. These are the factors that caused the exothermic reactions (3-9) and (3-10) *not* to go to completion. Instead, the atoms lodge in the best (most probable) compromise, *taking all sorts of randomness into account*. This always leads to an equilibrium situation in which there is something of everything present, the relative amounts being determined by the specific energy and positional randomness contributions.

(g) WEIGHTING THE ENERGY RANDOMNESS

As we consider reactions (3-8), (3-9) and (3-10), our experimental results reveal the significant effect of temperature. At room temperature, reaction (3-8), with its 56 kcal/mole heat of reaction, caused almost complete formation of N_2H_4, whereas reaction (3-9) with its 20 kcal exothermicity, was incomplete. However, at 400°K, even reaction (3-9) was only 60 percent complete and reaction (3-10) barely took place at all. The relative importance of energy randomness compared with positional randomness decreases considerably as the temperature rises. The explanation can be based upon the "expense" of energy randomness. For example, if the average molecular translational energy is $(3/2)kT$, then the average translational energy per mole increases as T increases. In effect, our 56 kcal/mole become less and less effective as a means of increasing motional randomness as the average motional energy per mole becomes comparable to the heat of reaction. To be explicit, $(3/2)kT$ erg per molecule corresponds to $(3/2)RT$ kcal per mole, and we can make the following comparisons:

at $T = 300°K$ (room temperature) $\frac{3}{2}RT = 0.9$ kcal
at $T = 400°K$ $\frac{3}{2}RT = 1.2$ kcal

Table 3-4 shows the significance of these comparisons to reactions (3-8), (3-9), and (3-10). At both 300°K and 400°K, the 56 kcal of heat released in reaction (3-8) imply a very large number of translational energy packets if each is considered to have an average of $(3/2)RT$ kcal per mole. For the other two reactions, however, the magnitude of $(3/2)RT$ is not so very large compared with ΔH. In fact, the ratio $\Delta H/(3/2)RT$ for reaction (3-10) at 300°K has nearly the same value as for reaction (3-9) at the higher temperature 400°K. This correlates with the fact that NF_2 reacts at 400°K to about the same extent as NO_2 at 300°K. This emphasizes what was said earlier: as temperature rises, the energy randomness becomes less and less important compared with positional randomness. The ratio $\Delta H/RT$ measures the importance of energy randomness. If the temperature is raised to the point that ΔH is small compared with RT, energy randomness will be quite unimportant. Then, dividing a few kilocalories up among a group of energy-rich molecules effects their energy-randomness about as much as dividing ten dollars among five millionaires effects their wealth.

Table 3-4 The Ratio of Heat of Reaction to Average Thermal Energy

	Reaction	ΔH (kcal/mole)	$-\Delta H/\frac{3}{2}RT$ 300°K	400°K
(3-8)	$2NH_2(g) = N_2H_4(g)$	−56	62	50
(3-9)	$2NF_2(g) = N_2F_4(g)$	−20	22	17
(3-10)	$2NO_2(g) = N_2O_4(g)$	−14	16	12

We see that the randomness to be gained by converting the potential energy change (ΔH) into energy of motion is measured by the relative magnitude of ΔH and the average energy per degree of freedom, which is determined by the temperature.

At low temperatures, the energy randomness will be quite important and equilibrium will favor the reaction in the exothermic direction (i.e., toward N_2H_4, N_2F_4, and N_2O_4 in our three examples). At high temperatures, energy randomness becomes less and less important, as $\Delta H/T$ becomes small. Then positional randomness will become dominant and the reaction may be favored in the endothermic direction (i.e., toward NH_2, NF_2, and NO_2). We shall expect to find $\Delta H/T$ an indicator of the temperature range at which this will begin to occur. In fact, $\Delta H/T$ is a measure of the energy randomness introduced by the heat of a reaction.

3-3 Randomness, energy change, and equilibrium

We have before us now all of the ideas needed to understand the factors that motivate spontaneous change. On a macroscopic scale, events move in the direction of increasing randomness. To be sure, the concept of randomness is more sophisticated than the simple positional randomness of our ping-pong ball gambling game shown in Figure 3-3. Energy must be partitioned, as well. Energy lodged in chaotic translational, vibrational, and rotational movements constitutes heat, and, in general, transfer of the potential energy of chemical bonds into heat increases the randomness. Thus, this energy randomization permits reactions to proceed in the exothermic direction. Positional randomness and molecular complexity also play their roles, however. Several molecular fragments can assume more positional arrangements than a single molecular cluster, even if the cluster has lower potential energy. The balance between all forms of randomness, including both reactants and products, determines the state of equilibrium.

We need to make all this quantitative. If randomness can be placed on a quantitative basis, we can predict reactions that might occur spontaneously—we can point the way towards equilibrium. We will do this in a rather laborious, but revealing derivation. The treatment does not adhere to the historical development of the subject—that would take us afield from chemistry, towards steam engines, and it would require quite a bit of mathematics. Again, we will favor the logical rather than the chronological, and the intuitively meaningful rather than the mathematically elegant. With apologies to Carnot and Clausius, we'll show in Chapter Four how economy and exercise lead to entropy, and how entropy leads to the Second Law of Thermodynamics.

A river tumbling down a mountain valley expends an immense amount of potential energy which is converted in a spendthrift way into the random motions of heat. As the water tumbles over rocks and splashes against canyon walls, the directed flow of the river becomes the disordered motion of heat: the water is warmed.

Because of the inspiration of a creative engineer, a series of dams is placed across the valley. Now, as the water makes its way down to the sea, it passes through a succession of turbines which convert much of its potential energy into electrical energy. This electrical energy can be converted into light, heat, chemical energy, or mechanical work, as man desires. These two ways in which melting snows in the mountains reach our eternal sea have very different consequences to the welfare of man. It behooves us to understand the work that can be performed (or lost) as spontaneous changes occur.

We shall see that any spontaneous process can perform work. The amount of work depends upon external conditions and it can approach a maximum quantity that is characteristic of the process. The difference between this theoretical maximum work and the actual amount performed in a real change will prove to be important to us. This difference is a quantitative measure of the "driving force" or "motivation" for the spontaneous change. It will lead us logically to a definition of entropy, and to an understanding of how randomness determines the state of equilibrium. We will explore first the work performed in a simple gas expansion, then that done in the discharge of an electrochemical cell.

four
maximum work, entropy and spontaneity

4-1 Expansion of a gas

Figure 4-1 shows the same changes of state that are pictured in Figure 3-1, but carried out in a new apparatus. Initially, N moles of chlorine gas are restrained in a 1-liter volume V_1 by the piston in the cylinder. The piston resists the gas pressure $P_1 = 24$ torr on its inner surface because of the weight M_1 on the suspension pan.

(a) EXPANSION INTO A VACUUM

In this apparatus we can easily bring about the expansion into a 3-liter volume (to give a final

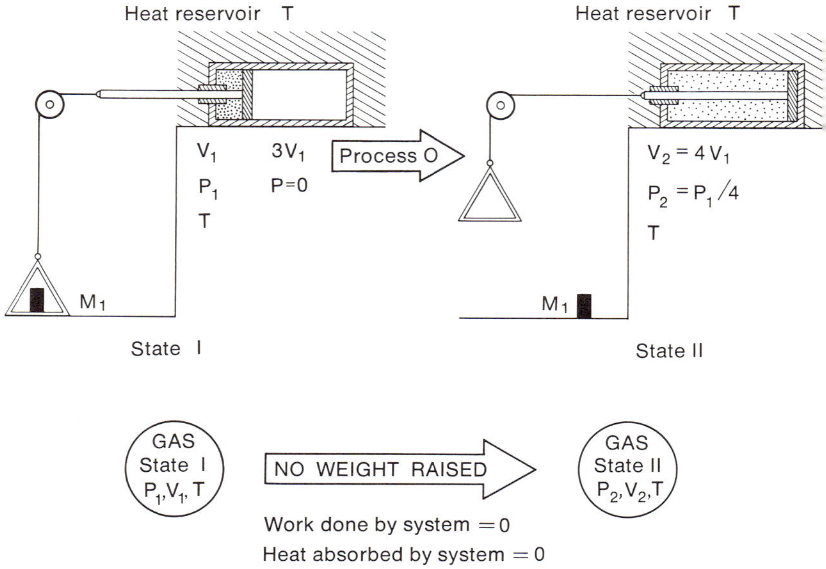

Figure 4-1 Expansion of a gas into a vacuum: No work.

volume of 4 l). If the weight M_1 is removed from the suspension pan, the piston is no longer restrained, and the gas pressure on the left-hand side of the piston pushes it rapidly to the end of the cylinder. There it is stopped by the wall, with a final volume of $V_2 = 4V_1$, and a final pressure of $P_2 = 6$ torr. Since the chlorine gas is at a low pressure, it behaves like an ideal gas, so there is no temperature effect. The entire expansion occurs at a constant temperature T.

The advantage of this experimental setup is that it is easy to see how much work is done as the gas goes from state I (N_1, P_1, V_1, T) to state II (N_1, P_2, V_2, T). In the process just considered, the mass M_1 was removed from the suspension pan, and except for the negligible mass of the suspension pan, no mass was lifted. The change occurred like that of the spontaneous and unrestrained flow of the river to the sea—no work was done.

(b) LET'S GET SOME WORK

Figure 4-2 shows another process by which this change could be carried out. Instead of merely removing the mass M_1 from the pan, it is replaced by a mass $\frac{1}{4}M_1$. The restraining force on the piston is now less than the gas pressure on the piston, so it moves to the right. The weight $\frac{1}{4}M_1$ is lifted and the gas continues to expand until the gas pressure is reduced to $\frac{1}{4}P_1$. At this pressure, balance is restored. Again, the gas pushing to the right exerts a force on the piston just equal to the force pulling the piston to the left due to the smaller mass on the pan.

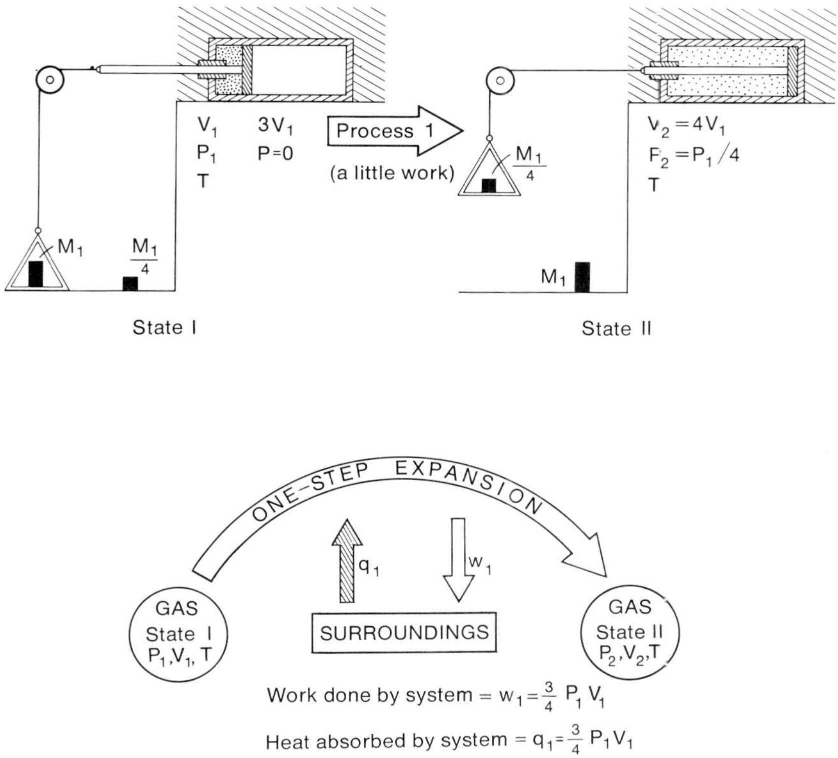

Figure 4-2 Expansion of a gas: Process 1, A little work.

We see that, as far as the gas is concerned, the change shown in Figure 4-2 is the same as that in Figure 4-1. The gas expanded from the same initial state I (N_1, P_1, V_1, T) to the same final state II (N_1, P_2, V_2, T). However, work was done as the weight was lifted. The gas performed useful work. We can call this work w_1 for the moment. The mass $\frac{1}{4}M_1$ is at an altitude h, and its potential energy is increased by w_1, where

$$w_1 = (\tfrac{1}{4}M_1)gh \tag{4-1}$$

There is, however, another way to express the work done. In the final state II, the balanced position of the piston shows that the weight $\frac{1}{4}M_1$ exerts just the same force on the piston as the pressure $P_2 = \frac{1}{4}P_1 = 6$ torr. As far as the piston is concerned, it might as well have the same pressure on both sides and no weight on the pan. Lifting the weight $\frac{1}{4}M_1$ is the same amount of work as if the piston had pushed, during its movement, against a constant pressure of $\frac{1}{4}P_1$. Hence, the work done in the process can be expressed in terms that relate directly to the actual change in state; the pressure and the volume.

If the work done is equal to the work of pushing back a constant pressure atmosphere, it is compression work. This we know about. In Chapter Two, this work was shown to be merely $P\Delta V$. (equation (2-16a)). It is work performed *by the system*, that is, by the gas pushing on the piston; we'll call it w_1.

$$w_1 = P\Delta V = \tfrac{1}{4}P_1(V_2 - V_1) = \tfrac{1}{4}P_1(4V_1 - V_1)$$

Hence,

$$w_1 = \tfrac{3}{4}(P_1V_1) \tag{4-2}$$

Compared with the process shown in Figure 4-1, we are much better off. In Figure 4-1 (process 0), no work was performed. In Figure 4-2 (process 1), a weight was lifted and work equivalent to $\tfrac{3}{4}(P_1V_1)$ was performed. The gas spontaneously made the same change of state, but in process 1 we got something out of it: the weight now has potential energy which can be released any time we wish to have a spot of energy to break a stone, cook an egg, or run an electric razor.

(c) STILL MORE WORK, PLEASE!

Even more work can be obtained in return for a little more attention. Consider process 2, shown in Figure 4-3. This time, we initially substitute for M_1 the mass $\tfrac{1}{2}M_1$. The gas pressure on the piston moves it to the right until balance is restored. This occurs when the pressure is halved and the volume doubled. An increment of work w_{2a} has been done, and we are in state Ia:

$$w_{2a} = P\Delta V = \tfrac{1}{2}P_1(V_{Ia} - V_I) = \tfrac{1}{2}P_1(2V_1 - V_1) = \tfrac{1}{2}P_1V_1 \tag{4-3}$$

At this juncture, the smaller weight $\tfrac{1}{4}M_1$ is substituted for $\tfrac{1}{2}M_1$, and the expansion proceeds the rest of the way to state II. Another increment of work w_{2b} has been done:

$$w_{2b} = P\Delta V = \tfrac{1}{4}P_1(V_2 - V_{Ia}) = \tfrac{1}{4}P_1(4V_1 - 2V_1) = \tfrac{1}{2}P_1V_1 \tag{4-4}$$

Now the total work done w_2 is simply the sum of w_{2a} and w_{2b}.

$$\begin{aligned} w_2 &= w_{2a} + w_{2b} = \tfrac{1}{2}P_1V_1 + \tfrac{1}{2}P_1V_1 \\ &= P_1V_1 \end{aligned} \tag{4-5}$$

In process 1 the system performed an amount of work equal to $\tfrac{3}{4}P_1V_1$. In process 2 the gas underwent the same expansion, but the work it performed went up to P_1V_1.

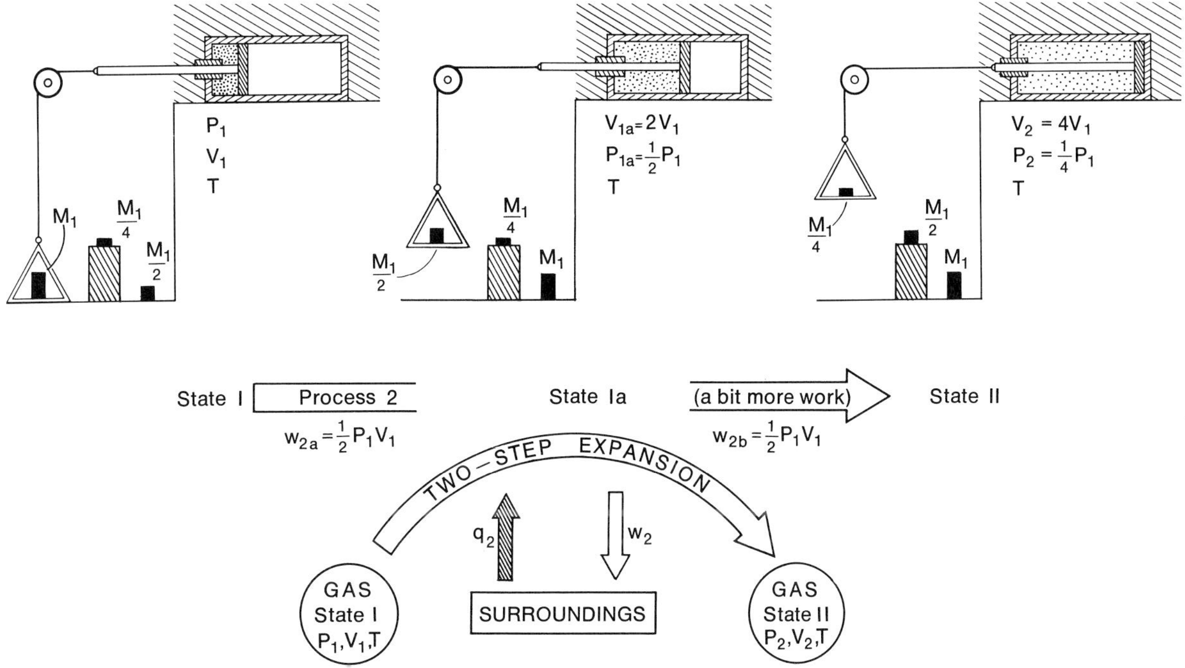

Work done by system $= w_{2a} + w_{2b} = P_1V_1$

Heat absorbed by system $= q_2 = P_1V_1$

Figure 4-3 Expansion of a gas: Process 2, More work.

Clearly this process could be modified even further to gain still more work. The first weight lifted could be $\frac{2}{3}M_1$, which could be replaced by $\frac{1}{2}M_1$, and finally by $\frac{1}{4}M_1$. The patience required for this step-wise process is repaid in additional work, as can be seen in a graphical presentation of the work done. Each step in each process is a constant pressure step and contributes an increment of work $P\Delta V$. This is readily seen to be the area in a plot of P against V, as shown in Figure 4-4. The work done in each process is represented by the shaded area. The shaded area in process 3 is plainly larger than that in process 2; it includes all of the shaded area in process 2 and more. More work is done.

Obviously we obtain more and more work as weight is taken off the pan in smaller and smaller increments. If maximum work is our aim (and why shouldn't it be—remember the creative engineer and his hydroelectric-dam project), we should carry the incremental process right to the limit of extremely small mass increments, so that at every instant the expanding gas lifts a mass that is only very slightly under that needed for balance. Such small mass increments are called "differential masses" and are abbreviated dM. The implication is that every expansion occurs at a constant pressure only dP (differential of pressure) below the equilibrium or balance pressure. Of course, the volume then changes in each step by only a differential amount dV. As the differential gets smaller and smaller, the work increases, as shown in Figure 4-5, approaching, finally, the smooth curve on the right. In this limit, the summation of incremental contributions becomes a continuous process. Mathematicians have a special name, the integral, for such an infinite sum of infinitesimally small contributions. The maximum work is given by this limiting sum.

$$\text{work} = \int_{\mathrm{I}}^{\mathrm{II}} P\,dV \tag{4-6}$$

This esoteric equation is read: "work equals the integral of the product PdV from state I to state II." The word "integral" means the limiting sum when the change from state I to state II is accomplished in smaller and smaller incremental steps.*

This limiting sum can be expressed analytically. Since the pressure is always kept extremely close to the pressure of the gas, we can apply the ideal gas law throughout the change.

$$P = \frac{nRT}{V} \tag{4-7}$$

*Mathematical integration is discussed in Appendix A.

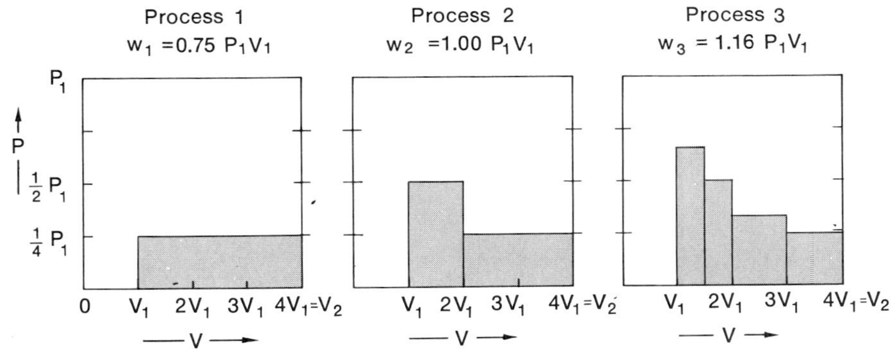

Figure 4-4 Graphical presentation of work: PΔV.

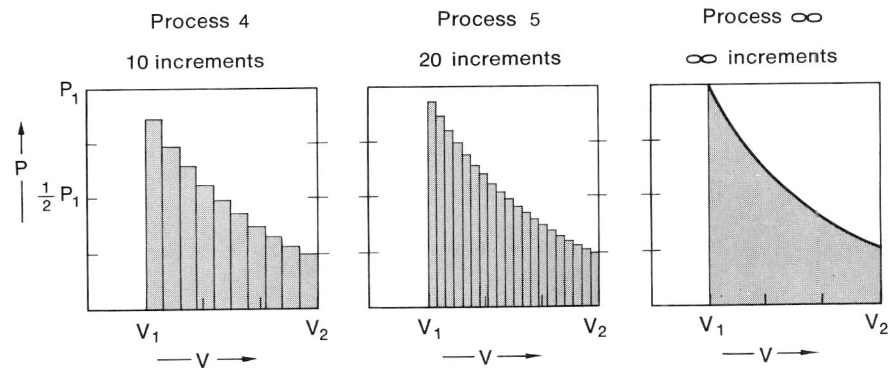

Figure 4-5 Expansion of a gas: Approach to maximum work.

where n is the number of moles of gas, R is the gas constant, and T is the absolute temperature.

Substituting (4-7) into (4-6) we obtain

$$\text{work} = \int_{\text{I}}^{\text{II}} \frac{nRT}{V} dV \qquad (4\text{-}8)$$

The summation in (4-8) can be simplified since the conditions do not permit either the number of moles or the temperature to change. If these factors (and R) do not change, they co not affect the integral except as a constant factor. Hence, (4-8) can be rewritten as

$$\text{work} = nRT \int_{\text{I}}^{\text{II}} \frac{dV}{V} \qquad (4\text{-}9)$$

In any elementary calculus class, this integral is one of the earliest considered. Its magnitude is relatively easy to calculate, since it is just a difference between two numbers fixed by the initial and final volumes V_1 and V_2. The difference in question is the natural logarithm of the volume in the final state minus the natural logarithm of the volume in the initial state.* Thus,

$$\text{work} = nRT \int_{\text{I}}^{\text{II}} \frac{dV}{V} = nRT \int_{V_1}^{V_2} \frac{dV}{V}$$

$$= nRT(\log_e V_2 - \log_e V_1) \qquad (4\text{-}10)$$

Since the difference between the logarithms of two numbers equals the logarithm of their quotient,

$$\text{work} = nRT \log_e \frac{V_2}{V_1} \qquad (4\text{-}11)$$

Logarithms to the base e are less convenient than logarithms to the base 10, so for computational purposes, (4-11) is usually written as

$$\text{work} = 2.303nRT \log_{10} \frac{V_2}{V_1} \qquad (4\text{-}12)$$

Substituting $V_2 = 4V_1$, (4-12) becomes

$$\begin{aligned} \text{work} = w_\infty &= 2.303nRT \log_{10} \frac{4V_1}{V_1} \\ &= 2.303nRT \log_{10} 4 \\ &= 2.303 \cdot 0.602nRT \\ &= 1.40nRT \end{aligned} \qquad (4\text{-}13)$$

To compare this result with our earlier processes, let us use the perfect gas relation once more, $P_1V_1 = nRT$. Thus,

$$w_\infty = \text{maximum work} = 1.40P_1V_1 \qquad (4\text{-}14)$$

Now we can sit back and assess the work done by the expanding gas in the various processes. Figure 4-6 shows a plot of the work as a function of the number of steps in the expansion. Both Figures 4-5 and 4-6 show that the process in which the pressure on the piston is almost balanced throughout the expansion gives the maximum work. That is, *maximum work is obtained when the system remains essentially at equilibrium throughout the process.*

It's too bad this last process is so tedious. Unfortunately, it takes an infinite time to expand a gas in an infinite number of steps.

*In Appendix A, the natural logarithm and the quantity e are defined, and the integral $\int dV/V$ is further discussed.

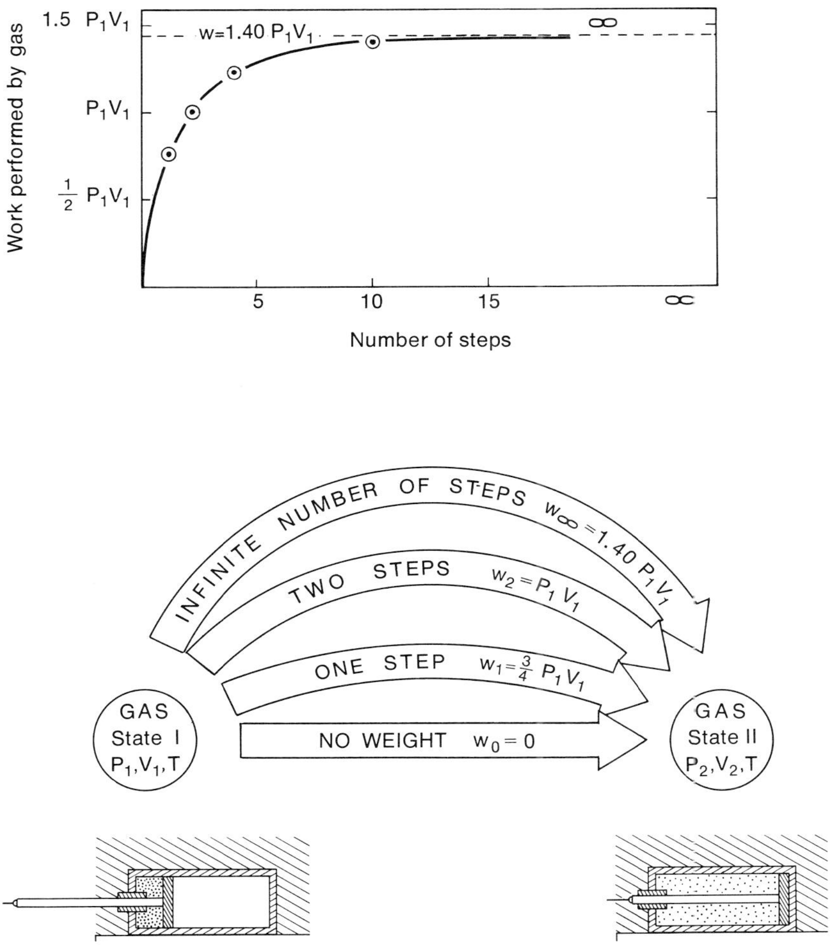

Figure 4-6 Work performed in a gas expansion.

(e) WHO PAID THE BILL?

In the several processes we have considered, all for the same change of state, only the work was discussed. It takes energy to do work—so who paid the energy bill? The answer is found in the First Law of Thermodynamics, the conservation law.

Our system is a quantity of gas in the state of a perfect gas; that is, a gas in which the molecules are so far apart that they do not interact noticeably. Such a gas expands at constant temperature without any energy change. Hence, the energy change ΔE during the change of state is zero:

$$\Delta E = E_2 - E_1 = 0 \qquad (4\text{-}15)$$

But also, the energy change must always equal the heat absorbed by the system, q, minus the work done by the system, w. That is,

$$\Delta E = q - w \tag{4-16}$$

and since $\Delta E = 0$,

$$q = w \tag{4-17}$$

So, for each of these processes, it was necessary to supply the gas with heat from a thermal reservoir.

So now we see who paid the bill. The surroundings (the thermal reservoir) had to give energy to the system so it could perform the appropriate amount of work. In each process, the amount of heat withdrawn equalled the amount of work performed.

(f) NOW LET'S GET THINGS BACK

Having studied this expansion, it is now time to restore the gas to its initial conditions. Quite apart from tidyness, there is a lesson to be learned.

To restore our system to state I, the gas must be compressed back into volume V_1. This change does not occur spontaneously—muscle must be applied to the piston. The weight on the pan must exceed the gas pressure in the cylinder to move the piston. Once again there are many ways we could do this. To get it over quickly, we could lift a weight up onto the pan equivalent to the final pressure P_1. That weight would surely pull the piston back, thereby recompressing the gas. The volume would decrease until pressure balance is restored, and this pressure balance would be reached when the volume is again V_1.

There are two steps involved. First, *we* do some work *w' on the weight M_1* as we lift it. Then, as the pan lowers, compressing the gas, *the weight* performs this same amount of work *w' on the system.* We'll discuss the restoration of the gas in terms of w', the work that *we* perform *on* the system (we and the weight are part of the surroundings). However, we can always re-express the work transfer in terms of w, the work done *by* the system, since $w = -w'$. In other words, *any* communication of work between a system and its surroundings can be expressed equally well in terms of w, the work done by the system, or in terms of w', the work done by the surroundings. These must always be equal in magnitude and opposite in sign: $w = -w'$.

$$\begin{matrix}\text{work we performed to} \\ \text{recompress the gas}\end{matrix} = \begin{matrix}\text{work performed to lift} \\ M_1 \text{ up onto the pan}\end{matrix} = w_1'$$

$$w_1' = P_1(4V_1 - V_1) = 3P_1V_1 \tag{4-18}$$

What's this? We have to perform an amount of work $3P_1V_1$ to go from state II to state I, when the gas would perform, at the very most, only $1.4P_1V_1$ of work as it went from state I to state II! Ouch, that was expensive.

(g) THERE MUST BE A CHEAPER WAY!

Indeed there is. Our technique in carrying out the expansion gives us the clue. Suppose the compression is also carried out in steps. Instead of placing the mass M_1 on the pan at the top of its rise, we could use $\frac{1}{2}M_1$. This mass would compress the gas only until its pressure was doubled, when the volume would be $2V_1$. Then the full weight M_1 could be used to finish the job. The work we have done is now the sum of two increments:

$$w_2' = \tfrac{1}{2}P_1(4V_1 - 2V_1) + P_1(2V_1 - V_1) = 2P_1V_1$$

These two processes are shown in Figure 4-7. Plainly we are in the same sort of game as that pictured in Figures 4-1–4-3, except that now *we* are doing all the work. If we are to recompress the gas with still less effort, we shall again have to carry out the change in many small steps. Leaping to the obvious conclusion, the more steps employed, the less work we need do. The minimum work w_∞' will be expended when an infinite number of steps is employed, always keeping the mass on the pan no more than an infinitesimal amount above the gas pressure in the cylinder. If we return to a consideration of w_∞, work done by the system, it can be expressed again as an integral, except that now the volume is $4V_1$ in the initial state and V_1 in the final state.

$$\begin{aligned} w_\infty &= \int_{V=4V_1}^{V=V_1} P dV \\ &= nRT[\log_e V_1 - \log_e(4V_1)] && \textit{(4-19)} \\ &= -nRT[\log_e(4V_1) - \log_e V_1] && \textit{(4-20)} \\ &= -2.303nRT \log_{10} \frac{4V_1}{V_1} \\ &= -1.40nRT = -1.40P_1V_1 && \textit{(4-21)} \end{aligned}$$

Expression (4-21) is the same as (4-13) and (4-14), except that the algebraic sign is minus. This means negative work is done by the gas; that is, the *surroundings* must perform work on the gas. Therefore, our final result is as follows:

$$\begin{aligned} \text{minimum work done on the gas during recompression} &= w_\infty' \\ &= -w_\infty \\ &= 1.40P_1V_1 && \textit{(4-22)} \end{aligned}$$

Unfortunately, history repeats itself. To perform a minimum amount of work during the compression, an infinite number of steps are needed and the process will take forever (literally!).

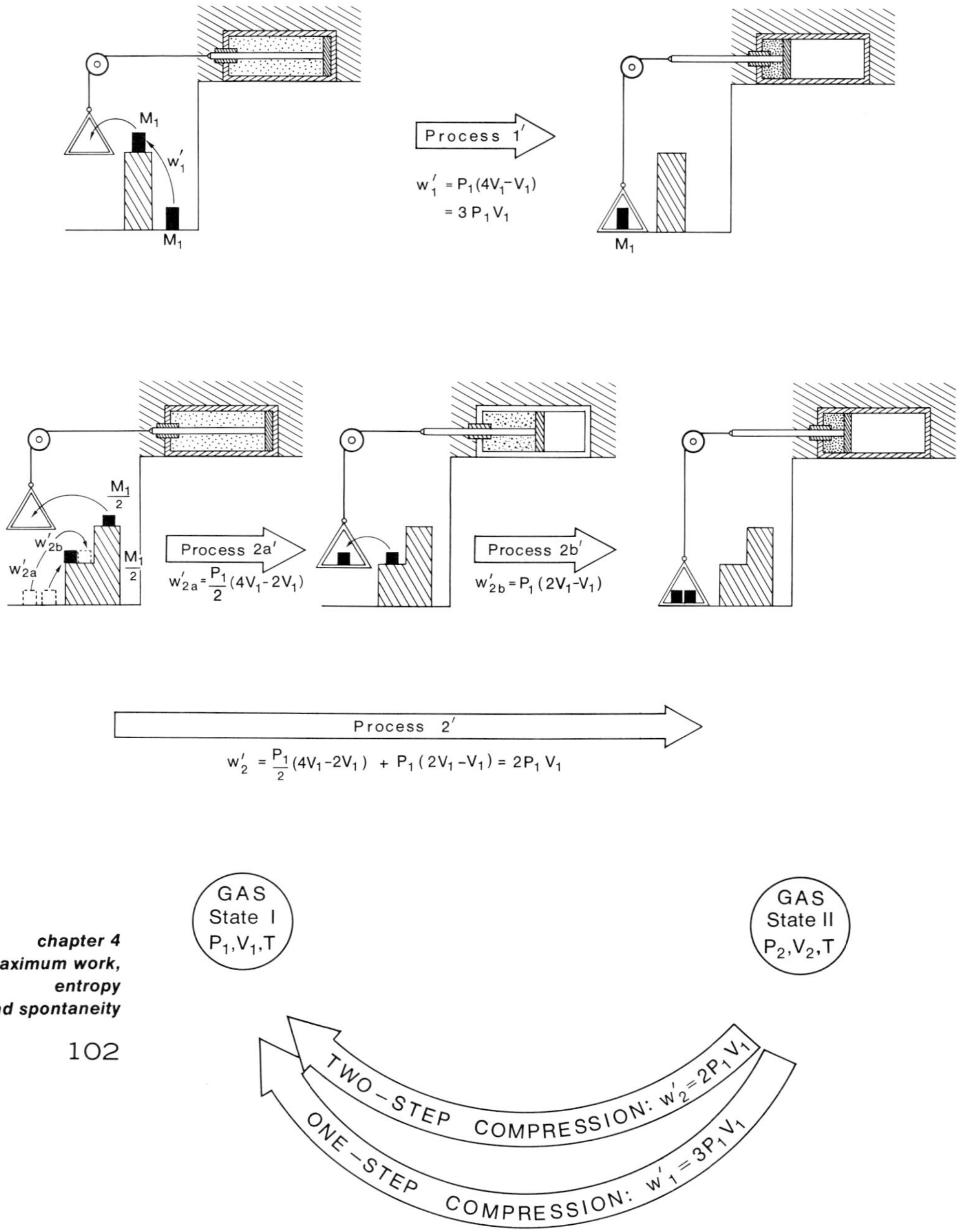

Figure 4-7 Getting the system restored with a minimum amount of effort.

(h) NOW WHO PAID THE BILL?

Just as before, the energy of the perfect gas is constant because the compression occurs at constant temperature. By the First Law of Thermodynamics,

$$\Delta E = q - w = 0$$

so

$$q = w$$

As always, w is the work done by the system, which is given by (4-21)

$$q = w_\infty = -1.40 P_1 V_1 \tag{4-23}$$

Equation (4-23) says that negative heat is absorbed by the gas. In other words, the gas releases heat into the thermal reservoir. If one of the less efficient processes is used (as in Fig. 4-7), the heat transferred into the reservoir is even greater. Who pays the bill? Whoever does the work of compression! Hey, that's us!

Table 4.1 shows a box score of the energy-moving: the transfer of heat between the reservoir and the gas, as well as the athletics with the weights. In each case, the heat absorbed and the work done by the surroundings are just the negative of the corresponding values for the system. With this table, we can see at a glance the net effect of carrying out the expansion and then the compression by any combination of processes. For example, suppose the expansion is carried out in a single step (process 1), producing work $\frac{3}{4}PV$, and the compression *also* in a single step (process 1′), requiring work $3PV$. Table 4.1 shows that the net work done by the gas is $\frac{3}{4}PV - 3PV = -\frac{9}{4}PV$, and the heat absorbed by the gas is $\frac{3}{4}PV - 3PV = -\frac{9}{4}PV$.

Table 4-1 Summary View of Energy Transferred In and Out of the Gas

Number of Steps	w, Work Done by the Gas	q, Heat Absorbed by the Gas
Expansion		
0 (no weight)	0	0
1	$0.75PV$	$0.75PV$
2	PV	PV
4	$1.16PV$	$1.16PV$
∞	$1.40PV$	$1.40PV$
Compression	**($w = -w'$)**	**($q = -q'$)**
1′	$-3PV$	$-3PV$
2′	$-2PV$	$-2PV$
4′	$-1.67PV$	$-1.67PV$
∞'	$-1.40PV$	$-1.40PV$

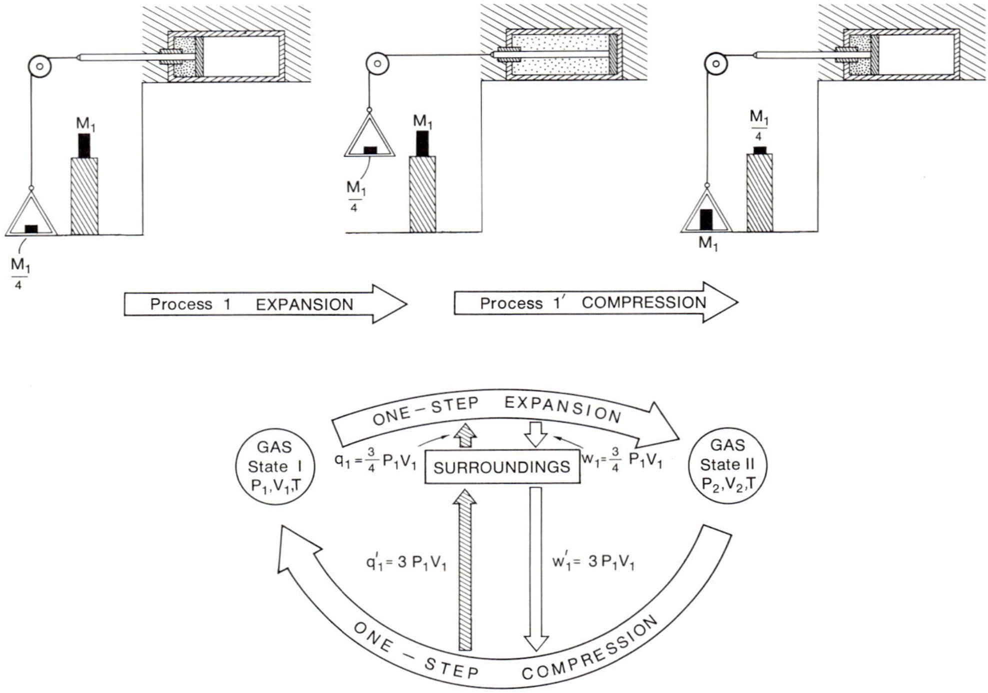

Net work: $3\,P_1V_1 - \frac{3}{4}P_1V_1 = 2\frac{1}{4}P_1V_1$ done by surroundings

Net heat: $3\,P_1V_1 - \frac{3}{4}P_1V_1 = 2\frac{1}{4}P_1V_1$ appears in surroundings

Figure 4-8 A possible expansion–compression cycle.

The negative signs mean that work was performed *on* the gas and heat was transferred *into* the surroundings. So the overall result is as shown in Figure 4-8. The gas has undergone a cyclic process —it has gone through a change but it has been exactly restored to its initial state. The surroundings, however, have been permanently changed. As shown by the arrows, the surroundings performed an amount of work $\frac{9}{4}PV$, *all of which ended up in the surroundings as heat.* The net effect of the cyclic process is that work (ordered motion) was transformed into heat (disordered motion).

Now we are mainly interested in the expansion process. We can focus on it by considering the *hypothetical* cyclic process in which we carry out the expansion in a fixed number of steps (as in the top half of Table 4-1) and *then recompress in the cheapest possible way.* This last part makes the process hypothetical, because it would take an infinitely long time to complete the compression step. Nevertheless, it lets us look more closely at the nature of the expansion process.

For example, consider the expansion by process 2 and recompression by process ∞'. Figure 4-8 is changed to Figure 4-9.

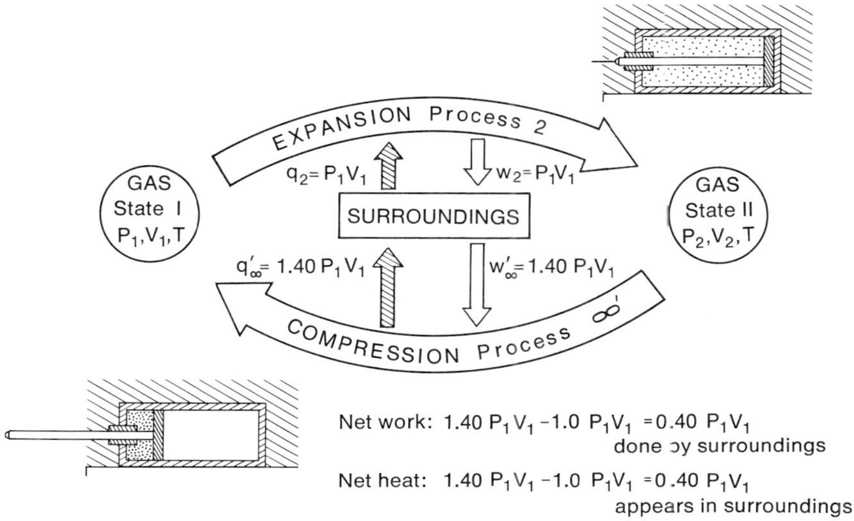

Figure 4-9 A more efficient expansion–compression cycle.

Again the gas is left in its initial state after the cyclic process. The route followed this time, however, requires that the surroundings perform a smaller net amount of work 0.4*PV,* all of which is converted into heat.

Clearly, if we wish to complete the cyclic process with a minimum waste of ordered work degenerated into disordered heat, we should carry out the expansion in as many steps as time permits. In fact, the most efficient process is that shown in Figure 4-10.

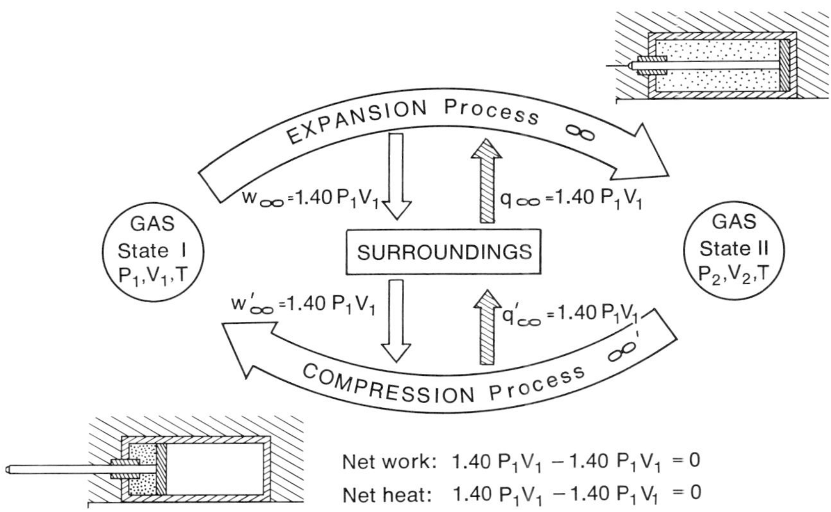

Figure 4-10 The most efficient possible expansion–compression cycle.

Here is the lazy man's perfect process! In the first step, the surroundings receive an amount of work 1.40*PV*, which just equals that expended in the second step. Not only is the gas carried through a cyclic process, but also the surroundings are unchanged at the end. During this tedious incremental process, we keep the system at pressure equilibrium throughout the expansion, and then maximum work is extracted from the gas as the change occurs. Then with an equally tedious compression, we can exactly reverse the changes *both in the* system and in the surroundings.

An equilibrium expansion is exactly reversible. Any nonequilibrium expansion process requires that work be converted into heat to restore the system to its initial state. We shall see that this important result applies to chemical reactions, as well as to gas expansions.

4-2 A chemical reaction in an electrochemical cell

When a piece of metallic zinc is immersed in an aqueous solution of silver nitrate, chemical reaction occurs. The zinc rod dissolves and bright silvery needles of metallic silver grow on the surface of the rod, as shown in Figure 4-11. Chemical analysis shows that the reaction is

$$Zn(m) + 2Ag^{+}(aq) \rightarrow Zn^{+2}(aq) + 2Ag(m) \qquad (4\text{-}24)$$

Like the river flowing unimpeded to the sea, the spontaneous change occurs in the beaker without the production of work. Though energy is released as the reaction occurs (it is an exothermic reaction), the energy merely heats up the solution and the surroundings.

We are reminded of the creative engineer, mentioned at the beginning of this chapter, who showed us how to convert the mechanical work in a waterfall to electrical energy. This electrical energy can be used directly to run motors, light lamps, or heat ovens, but it can also be stored in an electrochemical cell—then, when needed, the electrical energy can be recovered in the form of useful work. Why don't we construct an electrochemical cell from zinc and silver nitrate solution? Then our reaction (4-24) could do work for us. We shall see how to do this, modelling our approach after that of extracting work from a gas expansion.

(a) THE OPERATION OF AN ELECTROCHEMICAL CELL

Figure 4-12 shows a zinc–silver electrochemical cell. On the left-hand side, the zinc rod is immersed in 1.0 M zinc nitrate solution, while on the right-hand side, a silver rod is immersed in 1.0 M silver nitrate solution.* Between the two beakers is a porous, ceramic wall to establish electrical continuity between the two solutions. Externally, the two metal rods are connected through a switch S and a current meter (ammeter) A.

When S is closed, a current flows. Electrons leave the zinc rod, pass through the ammeter, and enter the silver rod. We note that changes occur in the solutions. On the left-hand side, zinc metal dissolves to form Zn^{+2} ions. This is the source of the electrons that flow out of the zinc rod. On left-hand side,

$$Zn(m) \rightarrow Zn^{+2}(aq) + 2e^- \qquad (4\text{-}25)$$

On the right-hand side, silver metal is plated out, consuming both the Ag^+ ions and the electrons that flow into the silver rod from the external circuit:

$$Ag^+(aq) + e^- \rightarrow Ag(m) \qquad (4\text{-}26)$$

Since the number of electrons leaving the zinc rod determines the amount of reaction (4-26) that can occur, we see that two moles of Ag(m) are formed for every mole of Zn(m) dissolved. The overall chemistry of the cell is

$$\begin{array}{r} Zn(m) \rightarrow Zn^{+2}(aq) + 2e^- \\ 2Ag^+(aq) + 2e^- \rightarrow 2Ag(m) \\ \hline Zn(m) + 2Ag^+(aq) \rightarrow Zn^{+2}(aq) + 2Ag(m) \end{array} \qquad (4\text{-}24)$$

This is the same reaction that occurred in the beaker shown in Figure 4-11. Now we can study this reaction under controlled conditions.

For example, we can measure the current flow. As operated in Figure 4-12, the factors that determine the current are the voltage generated by the cell $\mathscr{E}$, 1.56 volts, and the internal resistance of the cell R, which is 2.0 ohms. The current I is given by

$$\begin{aligned} I\text{ (amperes)} &= \frac{\mathscr{E}\text{ (volts)}}{R\text{ (ohms)}} \\ &= \frac{1.56}{2.0} \\ &= 0.78\text{ A} \end{aligned} \qquad (4\text{-}27)$$

One ampere is one coulomb of charge per second; so, if the switch were left closed for five minutes (300 sec), the charge that would leave the zinc rod and enter the silver rod is

$$\left(0.78\,\frac{\text{coulombs}}{\text{sec}}\right) \cdot (300\text{ sec}) = 234\text{ coulombs} \qquad (4\text{-}28)$$

*The choice of concentration is, of course, arbitrary. The use of 1.0 M solutions gives a standard set of conditions commonly used as a reference point.

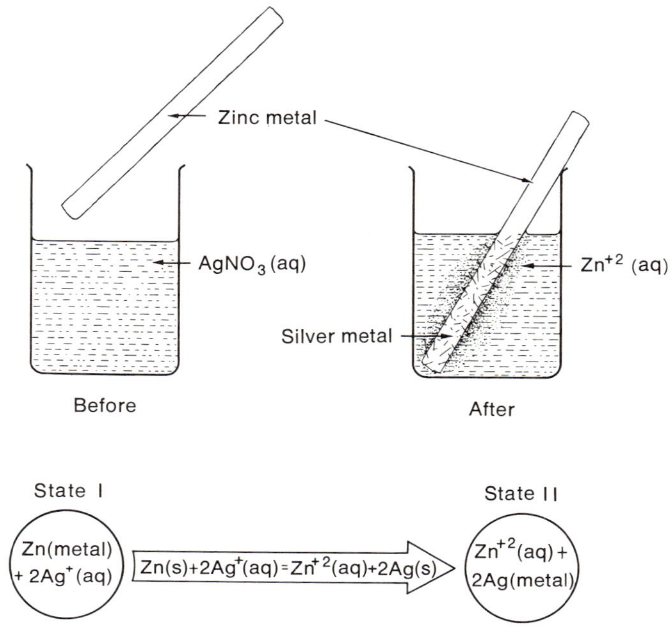

Figure 4-11 *A spontaneous chemical change taking place in a beaker.*

As chemists, though, we'd like to express this in moles, so we divide 234 by the number of coulombs per mole of electrons, 96,500. Then, the moles of charge moved in 300 seconds is

$$\frac{234}{96{,}500} = 2.4 \cdot 10^{-3} \text{ moles of electrons} \qquad \textit{(4-29)}$$

We are now ready to calculate the energy dissipation. The power (energy per unit time) generated by a current I in a resistance R is I^2R. The energy released is the power multiplied by the time for which the switch is closed. Hence, the energy effect is

$$\begin{aligned} \text{energy} &= (\text{power}) \cdot (\text{time}) \\ &= (I^2R) \cdot (t) \\ &= (0.78\ \frac{\text{coulombs}}{\text{sec}})^2 \cdot (2.0 \text{ ohms}) \cdot (300 \text{ sec}) \\ &= 365 \text{ joules} \\ &= (365 \text{ joules}) \cdot (0.239\ \frac{\text{calories}}{\text{joule}}) = 87 \text{ calories} \qquad \textit{(4-30)} \end{aligned}$$

Now we can assess the changes that occur during the 300 seconds that S is closed:

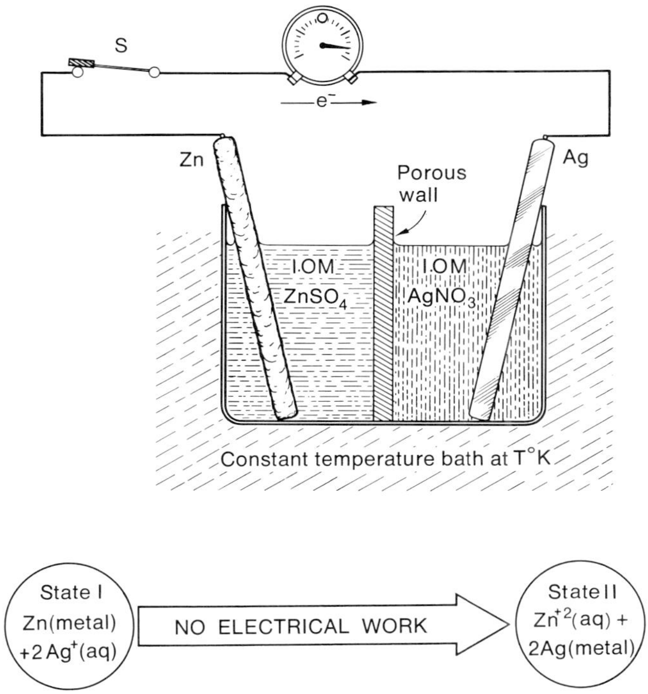

Figure 4-12 A chemical change in an electrochemical cell: No useful work.

(i) $2.4 \cdot 10^{-3}$ moles of electrons moved from the zinc rod to the silver rod.

(ii) $1.2 \cdot 10^{-3}$ moles of Zn metal dissolved.

(iii) $2.4 \cdot 10^{-3}$ moles of Ag metal were deposited.

(iv) 87 calories of heat were liberated in the internal resistance of the electrochemical cell, heating the solution.

Unfortunately, this is not useful work, any more than the heating of the solution in the beaker of Figure 4-11 was useful work. We wasted the available energy. *No useful work was performed* even though the reaction (4-24) was carried out in an electrochemical cell. The situation can be likened to the expansion carried out in Figure 4-1, if the piston has a bit of friction. No weight would be lifted (so no useful work would be done), but the friction would generate non-useful heat, just as the internal cell resistance does. We have yet to extract work from the electrochemical cell.

(b) LET'S GET SOME WORK

Figure 4-13(a) shows again our zinc–silver cell, but this time with the current passing through the windings of a motor. When current flows, the armature of the motor turns and useful work is performed.

To calculate the work output, we will take a simplified view of the motor by representing it as an equivalent resistance R_m, as pictured in Figure 4-13(b). We will then equate the heat released in R_m, due to the flow of current, to the useful work performed by the motor M.

Suppose the magnitude of the resistance R_m is 8 ohms. Then the total resistance will be the sum of R_m plus the internal cell resistance, $8 + 2 = 10$ ohms. Consequently, the current will be

$$I = \frac{\mathscr{E}}{R} = \frac{1.56}{10} = 0.156 \text{ amperes}$$

To obtain the same charge movement, $2.4 \cdot 10^{-3}$ moles of electrons, we would have to leave the switch closed for 25 minutes (1500 sec). During this time, heat would be generated both in the resistance (useful work), and in the cell (not useful work). Useful heat generated in R_m is:

$$\begin{aligned} \text{energy} &= (I^2 R_m)\,(\text{time}) \\ &= (1.56 \cdot 10^{-1})^2(8)(1500) \\ &= 292 \text{ joules} \\ &= 70 \text{ calories} \end{aligned}$$

Non-useful heat generated in the cell is:

$$\begin{aligned} \text{energy} &= (I^2 R)\,(\text{time}) \\ &= (1.56 \cdot 10^{-1})^2(2)(1500) \\ &= 73 \text{ joules} \\ &= 17 \text{ calories} \end{aligned}$$

We see that the same amount of heat is released, 87 calories, but now 80 percent of it is liberated in a useful form in the motor. Only 20 percent, or 17 calories, is dissipated in the electrochemical cell.

(c) STILL MORE WORK, PLEASE

Suppose a more powerful motor is used in Figure 4-13: let us say, a motor with an equivalent resistance of 38 ohms. Now the total resistance, $38 + 2 = 40$ ohms, will limit the current to $3.9 \cdot 10^{-2}$ amperes, so it will take 100 minutes (6000 sec) for $2.4 \cdot 10^{-3}$ moles

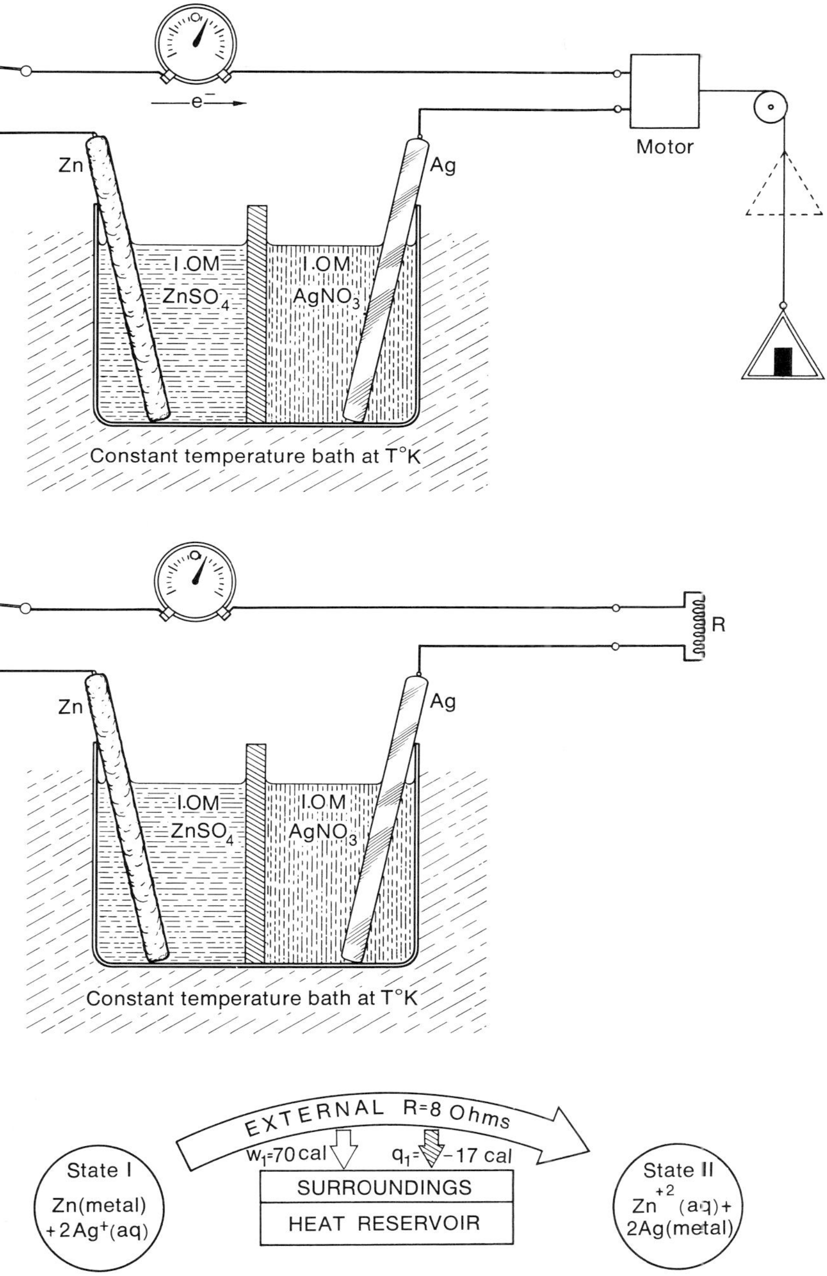

Figure 4-13 A chemical change in an electrochemical cell: Process 1, A little work.

of electrons to flow through the motor. The useful heat generated in R_m will now be:

$$\begin{aligned} \text{energy} &= (I^2R)\ (\text{time}) \\ &= (3.9 \cdot 10^{-2})^2(38)(6 \cdot 10^3) \\ &= 347 \text{ joules} \\ &= 83 \text{ calories} \end{aligned}$$

The non-useful heat generated in the cell will be:

$$\begin{aligned} \text{energy} &= (I^2R)\ (\text{time}) \\ &= (3.9 \cdot 10^{-2})^2(2)(6 \cdot 10^3) \\ &= 18 \text{ joules} \\ &= 4 \text{ calories} \end{aligned}$$

Again, the same amount of heat is released, but now 95 percent is obtained in useful heat! The only problem is that now it takes $1\frac{2}{3}$ hours to extract the energy.

(d) MAXIMUM WORK

Plainly we can get as high a fraction of available work as we desire by merely increasing the magnitude of R_m relative to the intrinsic cell resistance. The only problem is that to approach closer and closer to the maximum possible work takes a longer and longer time. For example, if $R_m = 198$ ohms, then 99.0 percent of the energy is extracted as useful heat, but the current must flow $8\frac{1}{3}$ hours to give us the 86 calories.

It is possible to extract work from a spontaneous chemical change (via an electrochemical cell), but just as in the expansion of a gas, to obtain maximum work, the work must be removed at an infinitesimally slow rate.

We have considered the conversion of electrical work into heat as equivalent to its conversion into mechanical work via a motor. An alternate way to extract the available energy usefully is to transfer it to a second electrochemical cell, by charging that cell. Of course, the cell to be charged must itself generate a voltage below that of the discharging cell so that the current flows in the right direction. Whether driving a motor or charging another cell, maximum work is obtained from a chemical reaction if the reaction occurs in a cell working against an external voltage lower than its own by no more than a differential amount.

We can obtain an analytical expression for the maximum useful work that can be extracted from a chemical reaction. Suppose we wish to extract maximum work from n moles of electrons (charge per electron $= e_0$) taken from a cell with constant voltage $\mathscr{E}$. The current I is

$$I = \frac{\mathscr{E}}{R} \qquad (4\text{-}27)$$

but also

$$I = \frac{\text{amount of charge moved}}{\text{time}} = \frac{ne_0}{t} \qquad (4\text{-}31)$$

Equating (4-27) to (4-31) and solving for *t,* we obtain

$$t = ne_0\frac{R}{\mathscr{E}}$$

$$\text{energy} = [I^2R](t) \qquad (4\text{-}30)$$

$$= \left[\left(\frac{\mathscr{E}}{R}\right)^2 R\right]\left(ne_0\frac{R}{\mathscr{E}}\right) = ne_0\mathscr{E} \qquad (4\text{-}32)$$

The available energy is fixed by the cell voltage $\mathscr{E}$ and *n,* the number of moles of charge extracted from the cell.

(e) NOW LET'S GET THINGS BACK

In each of the uses of the cell, we have considered the movement of the same amount of charge, $2.4 \cdot 10^{-3}$ moles of electrons. According to the overall chemistry of the cell, we must have deposited this same number of moles, $2.4 \cdot 10^{-3}$, of metallic silver, and dissolved half as many moles of metallic zinc.

$$\text{Zn(metal)} \rightarrow \text{Zn}^{+2}\text{(aq)} + 2\text{e}^-$$
$$2\text{Ag}^+\text{(aq)} + 2\text{e}^- \rightarrow 2\text{Ag(metal)}$$

The net reaction is

$$\text{Zn(metal)} + 2\text{Ag}^+\text{(aq)} \rightarrow \text{Zn}^{+2}\text{(aq)} + 2\text{Ag(metal)} \qquad (4\text{-}24)$$

Therefore, to restore the cell, the current through the cell must be reversed, reversing the chemistry of (4-24). This must be continued until $2.4 \cdot 10^{-3}$ moles of electrons have passed through in this reverse direction, redepositing $1.2 \cdot 10^{-3}$ moles of zinc and redissolving $2.4 \cdot 10^{-3}$ moles of silver. This reversal can be accomplished with another electrochemical cell of higher voltage, as shown in Figure 4-14. If, for example, we use an automobile battery with a voltage of 6.0 volts and an internal resistance of, say, 6 ohms, a large current will flow and our cell will be quickly restored. The net voltage available will be the difference between the two voltages, $6.0 - 1.56 = 4.44$ volts, since they are opposed to each other. The current that will flow (0.56 amperes) is determined by this net voltage and the sum of the two battery resistances, $6 + 2 = 8$ ohms. By our now-familiar calculations, we find

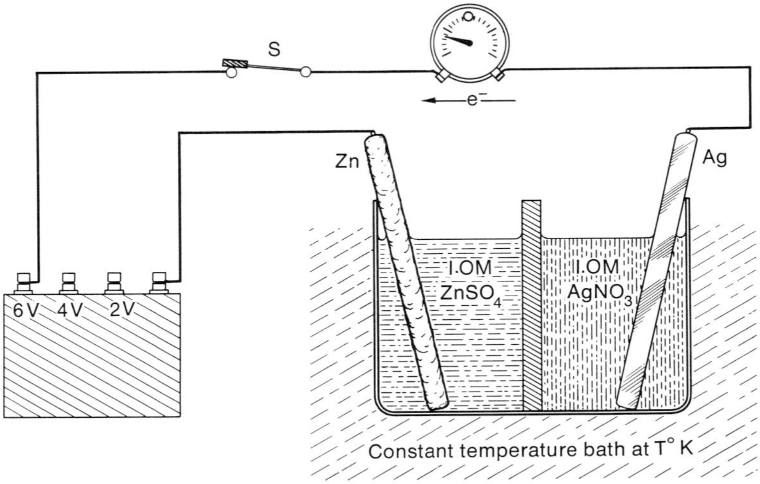

Net $\mathcal{E} = 6.0 - 1.56 = 4.44$ volts

Current $I = \frac{\mathcal{E}}{R} = \frac{4.44}{8} = 0.56$ amperes

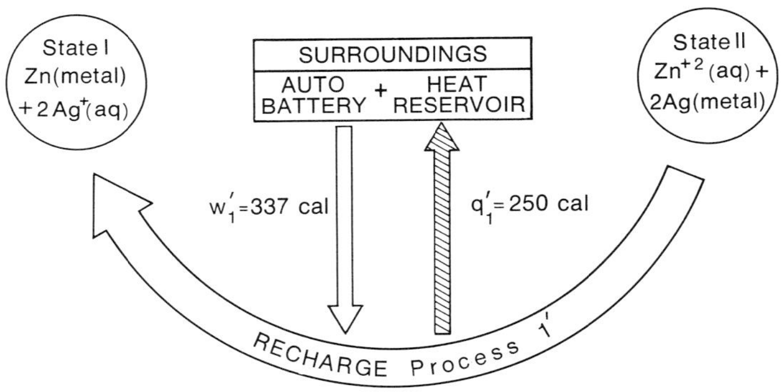

Figure 4-14 Recharging the cell: Process 1'.

that this current must flow for 417 seconds (7 min), and during this time, non-useful heat will be generated in the two cells:

$$
\begin{aligned}
\text{energy (non-useful heat)} &= (I^2R)\,(\text{time}) \\
&= (0.56)^2(8)(417) \\
&= 1046 \text{ joules} \\
&= 250 \text{ calories}
\end{aligned}
$$

Again, ouch! That was expensive! Our cell is recharged—we restored its capacity to deliver 87 calories but we wasted 250 calories in useless heating to do so!

(f) THERE MUST BE A CHEAPER WAY!

Suppose we cut back the voltage used to recharge our cell. As shown in Figure 4-15, we might use only two of the cells of the automobile battery. This gives a voltage of 4.0 volts which, when set against the Zn–Ag cell, gives a net voltage of 2.44 volts. Now the current must flow for 570 seconds ($9\frac{1}{2}$ min), but less energy is wasted in heating the cells. The cell is recharged, but only 137 calories are dissipated.

An even more economical recharge is obtained with only one of the automobile battery cells. Now the net charging voltage is quite small, $2.0 - 1.56 = 0.44$ volts. Of course, that means the current will be much smaller, 0.11 amperes, and it will take quite a while to move the desired charge, $2.4 \cdot 10^{-3}$ moles of electrons.

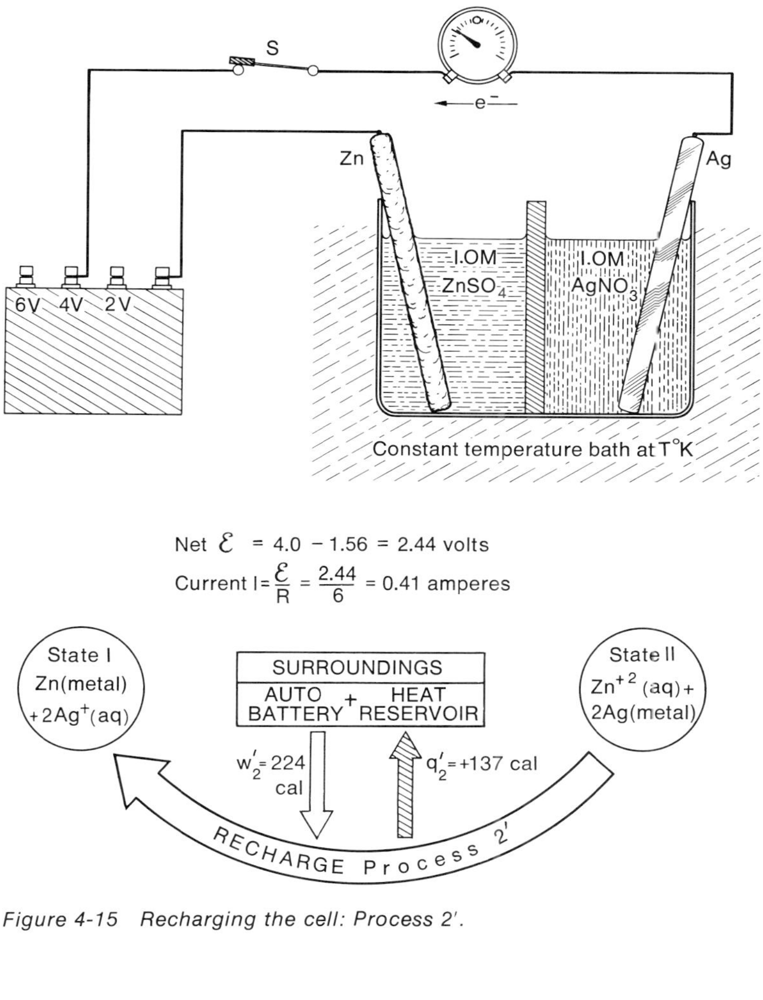

Figure 4-15 Recharging the cell: Process 2′.

The recharge time becomes $35\frac{1}{2}$ minutes (2125 sec), but only 25 calories of useless heat are produced.

Plainly, the way to recharge our cell with a minimum waste of energy is to recharge with a voltage source just barely above that of the Zn–Ag cell. This voltage source might be another cell, as pictured in Figures 4-14 and 4-15, or it might be a generator driven by a turbine in the engineer's hydraulic system. Whatever method is used, the more efficient our recharge is to be, the longer it will take to restore the Zn–Ag cell.

(g) SUMMARY: A CYCLIC USE OF AN ELECTROCHEMICAL CELL

In the same way that we carried a gas through cyclic expansion and recompression, we have now considered the cyclic discharge and recharge of an electrochemical cell. Table 4-2 shows the energy balance in each of the several processes considered.

Table 4-2 Summary View of Energy Transferred Out of and Into an Electrochemical Cell

Discharge				
Process	**External Resistance (volt)**	**Time Required (min)**	**Energy Wasted (cal)**	**Useful Work (cal)**
0	0	5	87	0
1	8	25	17	70
2	38	100	4	83
∞	∞	∞	0	87
Recharge				
Process	**Excess Voltage (volt)**	**Time Required (min)**	**Energy Wasted (cal)**	**Energy Stored (cal)**
1′	4.44	7	250	87
2′	2.44	9	137	87
3′	0.44	35	25	87
∞′	0.00	∞	0	87

Again we see that any combination of discharge and charge that takes a finite amount of time involves the waste of energy in undesired heating. For example, if the discharge requires 25 minutes, while 70 calories of work are obtained, 17 calories of energy are converted to unwanted heat. Then, if we recharge in 35 minutes, to replace this 87 calories, 25 calories of useless heat are produced. The cyclic process takes 60 minutes and its net effect in the surroundings is as shown in Figure 4-16.

In this case, during the discharge process, both work (70 calories of it) and heat (17 calories) are transferred to the surroundings. This energy is all drawn from the exothermic chemical

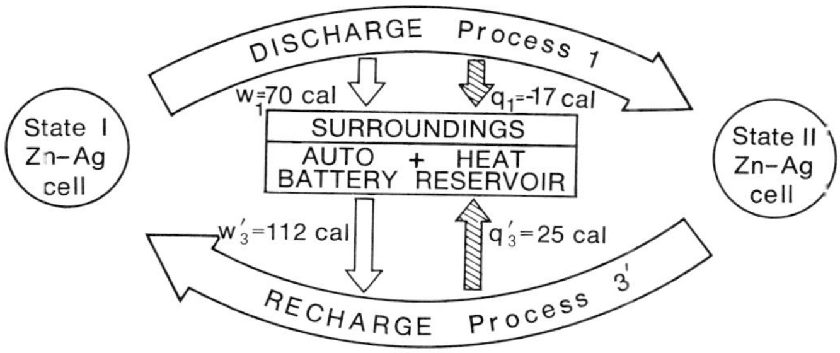

Net work: 112 − 70 = 42 cal done by surroundings

Net heat: 17 + 25 = 42 cal appears in surroundings

Figure 4-16 A possible discharge–recharge cycle.

reaction between metallic zinc and aqueous silver ions in reaction (4-24). Then, in the recharge process, the surroundings must do work on our cell: 112 calories by process 3′, of which 25 calories are expended as useless heating and also end up in the surroundings. The net effects are the following.

(i) Our cell is restored.

(ii) The surroundings received 70 calories of work and delivered 112 calories, a net expenditure of 42 calories of work by the surroundings.

(iii) Seventeen calories of heat were generated during discharge and 25 more during recharge, a total of 42 calories all of which appear in the surroundings.

We see that, in the 60 minute cyclic process, the overall effect in the surroundings is that 42 calories of work are converted to heat.

Once again attention can be focussed on the discharge process by considering the hypothetical cyclic process in which the recharging process ∞′ is used, even though it would take an infinite time. Now the 25 minute discharge process performs 70 calories of useful work out of the possible 87. It wastefully heats the surroundings with the remaining 17 calories. By the time the reversible recharging is completed, the surroundings have transformed a net 17 calories of work into heat. If 100 minutes are allowed for discharge, then the cyclic process degrades only 4 calories of work into heat. If the discharge could take an infinite amount of time, then and only then, could the cyclic discharge–recharge process exactly reverse the changes *both in the system and in the surroundings.* This would be accomplished by exactly balancing the cell voltage against an external voltage, so as to maintain electrical equilibrium at every moment in the cycle. *Such an equilibrium discharge of a cell is exactly reversible.* **Any non-equilibrium process,** *one that takes place spontaneously in a finite time,* **requires that work be converted into heat** *to restore the system to its initial state.*

(h) REVERSIBILITY

The result we have just obtained for the expansion of a perfect gas and for the discharge of an electrochemical cell proves to be true for any change of state. If a process can be found in which the change occurs entirely at equilibrium, the process is reversible. However, a reversible process can be closely approximated but never actually carried out, because it must take place infinitely slowly.

A change carried out reversibly can be restored in a reversible manner without any change in the surroundings. *In any real process* (one that takes place in a finite time), *the change can be restored only at the net expense of changing some work into heat.* Notice, however, that real processes are the ones that take place spontaneously. When the weight M_1 is removed from the pan, as in Figure 4-1 or Figure 4-2, the gas spontaneously pushes the piston to the right, lifting the pan. Then, no matter how carefully and slowly the gas is recompressed in the cyclic process, work will have been extracted from the surroundings and converted into heat in the surroundings. If a spontaneous chemical reaction is allowed to proceed in a beaker or in an electrochemical cell, as in Figures 4-11 or 4-12, it, too, can be reversed and the reactants restored. This can be done with another cell by reversing the current, as in Figure 4-14. However, as before, no matter how carefully and slowly the reaction is reversed in the cyclic process, work will have been extracted from the surroundings and converted into heat in the surroundings.

Thus, *the reversal of any spontaneous process always requires a change of work into heat.* The closer the process comes to an equilibrium process, the smaller will be the net change of work into heat when the process is reversed. At equilibrium, the energy degraded into heat will be a minimum. Later, we shall investigate the discrepancy between the actual heat q, in a spontaneous process, and the reversible heat q_{rev}, in a hypothetical, infinitely slow process. *We shall find this difference, $q_{rev} - q$, to be a measure of the tendency for a reaction to occur spontaneously.*

(i) THE SECOND LAW OF THERMODYNAMICS

Before exploring ($q_{rev} - q$), we should note that the gas expansion and the electrochemical cell discharge have revealed an important principle of nature. The principle is so widely applicable that it is called the Second Law of Thermodynamics. (The First Law states that "Energy is Conserved.") There are several equivalent statements of the Second Law; here are three of them.

The Second Law of Thermodynamics

(i) **All spontaneous processes are irreversible.** *(4-33)*

(ii) **After any spontaneous process, work must be converted to heat in order to restore the system to its initial state.** *(4-34)*

Since work is a manifestation of ordered energy and heat is disordered energy, the essence of the second law is that a system changes spontaneously in the direction of increasing randomness. This gives us a third statement of this many-faceted law.

(iii) **In a spontaneous process, randomness increases.** *(4-35)*

4-3 Entropy and probability

In Chapter Three it was concluded that a system tends to change spontaneously towards the most random arrangement Then, in Section 3-2(g), we surmised that the quantity $\Delta H/T$, enthalpy change divided by temperature, might be a measure of randomness. Now we add that a spontaneous process involves a change of work into heat.

It is time to introduce a symbol to represent this change in randomness and to attempt a quantitative connection between randomness and heat effects. The name scientists have chosen is **entropy** and its change is symbolized ΔS. We shall now show how entropy is related to heat effects.

(a) ENTROPY AND ITS RELATION TO PROBABILITY

The concept of entropy is connected to the intrinsic probability of a system. Simple ideas of probability tell us the nature of S. What properties should entropy display? These can be written down intuitively.

(i) *Entropy should be a state function.* An entropy change ΔS must be the difference between the initial and final state entropies S_1 and S_2, and must not be dependent upon the process by which the change occurs.

$$\Delta S = S_2 - S_1 \qquad (4\text{-}36)$$

(ii) *Entropy should be additive.* If a system is considered in parts, the total entropy S should be the sum of the entropies of the parts, say S' and S''.

$$S_{\text{total}} = S' + S'' \qquad (4\text{-}37)$$

In probability terms, the second condition is easily assured. Suppose a system can be considered to be made up of two parts. If the probability of finding part one in a given situation is designated W' and the probability of finding part two in a given situation is W'', then the overall probability W of finding *both* parts one and two in the given situations is the product of W' and W'',

$$W = W' \cdot W'' \quad (4\text{-}38)$$

The combination of (4-37) and (4-38) tells us that if entropy is a function of probability (that is, if $S = S(W)$), the mathematical form of $S(W)$ must be such that

$$S(W) = S(W') + S(W'')$$

or

$$S(W' \cdot W'') = S(W') + S(W'') \quad (4\text{-}39)$$

There is, in fact, only one function that satisfies (4-39), and a mathematician recognizes it at once. The logarithm has the property we desire.

$$\log(a \cdot b) = \log a + \log b$$

So if we are to have the mathematical property of (4-39), entropy must have a logarithmic dependence upon probability.

$$\boxed{S = \log W} \quad (4\text{-}40)$$

So, if $W = W'W''$,

$$S = \log(W'W'') = \log W' + \log W'' = S' + S''$$

(b) ENTROPY AND ITS RELATION TO HEAT IN A PERFECT GAS EXPANSION

The perfect gas expansion was first considered in Section 3-1(b) in terms of randomness. It was considered again in Section 4-1 with attention to the work and heat effects. Let us summarize the results.

Randomness: We found in Section 3-1(b) that "*probability leads to a situation that distributes molecules between two bulbs in proportion to their volumes.*" We can say this mathematically in terms of a proportionality:

W is proportional to V

or,

$$W = aV \quad (4\text{-}41)$$

where a is some proportionality constant we don't yet know.

If we wish to consider an expansion from state 1 with volume V_1, to state 2 with volume V_2, (4-41) tells us that

$$W_1 = aV_1$$
$$W_2 = aV_2$$

Each of these probabilities can be substituted into (4-40):

$$S_1 = \log W_1 = \log(aV_1) \tag{4-42}$$

$$S_2 = \log W_2 = \log(aV_2) \tag{4-43}$$

Now it is time to return to property (4-36). The *change* in entropy ΔS, in the expansion, must simply be

$$\begin{aligned} \Delta S &= S_2 - S_1 \\ &= (\log aV_2) - (\log aV_1) \\ &= \log \frac{aV_2}{aV_1} = \log \frac{V_2}{V_1} \end{aligned} \tag{4-44}$$

Oh, wonder of wonders! Expression (4-44) is immediately reminiscent of the maximum work that can be obtained in a reversible expansion, as given by (4-11):

$$\begin{matrix}\text{``reversible''} \\ \text{work done}\end{matrix} = \begin{matrix}\text{``reversible''} \\ \text{heat absorbed}\end{matrix} = nRT \log_e \frac{V_2}{V_1} \tag{4-11}$$

By using $\log_e$ in (4-44), we can substitute (4-44) into (4-11) and obtain

$$\begin{matrix}\text{reversible heat} \\ \text{absorbed}\end{matrix} = nRT \log_e \frac{V_2}{V_1} = nRT\Delta S$$

or, rearranging,

$$\Delta S = \frac{\text{reversible heat}}{nRT} \tag{4-45}$$

Expression (4-45) can be simplified by considering the entropy on a "per mole" basis; that is, by choosing $n = 1$.

$$\Delta S = \frac{\text{(reversible heat)}}{RT} = \frac{q_{\text{rev}}}{RT} \tag{4-46}$$

Expression (4-46) is usually written without the constant term $1/R$ (R = gas constant), but this is only a matter of traditional definition. If we had decided to define $S = R \log_e W$, then (4-46) would appear in the form usually seen:

$$\boxed{\Delta S = \frac{q_{\text{rev}}}{T}} \tag{4-47}$$

This can be taken as the definition of entropy. We have derived it for a simple process, the expansion of a perfect gas. The result (4-47) was obtained long ago in the nineteenth century, at a time when heat was treated as though it were a substance (called "caloric"), and before any connection with probability had been postulated. The usefulness of the concept of entropy has been verified in innumerable applications of chemistry and physics. Hence, the probability significance of entropy and the validity of the entropy concept throughout all chemical phenomena are now firmly based in experience and are accepted by all practicing scientists.

(c) ENTROPY AND THE SECOND LAW

We see that the probability of a given state is measured by the quantity called entropy. Since increasing randomness implies increasing probability, the second law (as expressed in form (iii), 4-35) can be restated in terms of entropy. Spontaneous changes occur in the direction that increases randomness, so entropy must also increase during any spontaneous change. Now the First and Second Laws of Thermodynamics can be stated in parallel and concise forms.

First Law

The energy of the universe is constant. *(4-48)*

Second Law

The entropy of the universe increases. *(4-49)*

4-4 Free energy and spontaneous change

We are on the verge of something great. The entropy, which has a simple microscopic significance, probability, can be measured through heat effects. Also, entropy tells us the direction of spontaneous change! If someone asks whether a given system will spontaneously undergo a certain change, the answer is to be found in the total effect on the entropy of the universe. If we can see that the combined entropies of the system and of the surroundings will increase, the answer is "Yes." If the combined entropies decrease, the change does not occur spontaneously. If the combined entropies remain constant, the system is at equilibrium. Obviously, to a chemist, entropy is as important as energy.

It would be more convenient, however, if this spontaneity criterion could be re-expressed in terms of properties of the system alone. The surroundings, after all, are the rest of the universe, which can present quite a bookkeeping problem. Fortunately this bookkeeping is readily done for changes that occur at constant temperature and constant pressure, the common laboratory situation of interest to us.

(a) REVERSIBLE HEAT AND SPONTANEOUS CHANGE

In Sections 4-1 and 4-2 we looked closely at two constant temperature processes. In each case we found that the difference between the maximum possible heat absorbed q_{rev}, and the actual heat absorbed q, is specially informative. For these constant temperature processes, the difference $(q_{\text{rev}} - q)$ measures the net change of work into heat when we try to restore the system reversibly. This net change of work into heat represents the "extent of irreversibility."

$$\text{extent of irreversibility} = q_{\text{rev}} - q \qquad (4\text{-}50)$$

We shall see that this quantity, expression (4-50), expresses the Second Law criterion for spontaneous change.

Therefore, let's return to Figure 4-6 which shows several of the ways in which the change in question can be brought about. First, consider process ∞, during which the maximum possible work $1.40PV$ was extracted from the expanding gas. This process is reversible, so the heat effect can be labeled q_{rev}. This heat effect immediately allows the specification of the entropy change for the *system*:

$$\text{heat absorbed} = 1.40PV = q_{\text{rev}} \qquad (4\text{-}51)$$

$$\text{entropy change} = \frac{q_{\text{rev}}}{T} \qquad (4\text{-}52)$$

As the system absorbed q_{rev}, the surroundings had to give up that same amount of heat. We can say, then, that the surroundings *absorbed* $(-q_{\text{rev}})$ calories of heat. Now the entropy change in the *surroundings* can be given:

$$\text{heat absorbed} = (-q_{\text{rev}}) \qquad (4\text{-}53)$$

$$\text{entropy change} = -\frac{q_{\text{rev}}}{T} \qquad (4\text{-}54)$$

The entropy change of the entire universe is the sum of (4-52) and (4-54). Adding these, we obtain

$$\begin{aligned}\Delta S_{\text{universe}} &= \Delta S_{\text{system}} + \Delta S_{\text{surroundings}}\\ &= \left(\frac{q_{\text{rev}}}{T}\right) + \left(-\frac{q_{\text{rev}}}{T}\right)\\ &= 0 \end{aligned} \qquad (4\text{-}55)$$

The reversible process causes no change in the entropy of the universe. The criterion for a spontaneous change is not met—entropy does not increase. Instead, it is exactly zero. Th s identifies an equilibrium process.

Now consider one of the irreversible paths in Figure 4-6. In any one of these processes, the work extracted will be less than the maximum possible. The heat effect q will be less than q_{rev}. Let's look again at the entropy changes in the system, the surroundings, and then their sum; the entropy change of the universe.

First consider the system. We already know its entropy change—entropy is a state function, $\Delta S = S_2 - S_1$, so it depends only upon the initial and final states. The entropy change of the system must be $q_{\text{rev}}/T = 1.40PV/T$, as given in (4-52), no matter how the expansion is carried out.

The entropy of the surroundings, however, is something else. As far as the heat reservoir is concerned, a certain amount of heat q was extracted at constant temperature. The surroundings (the reservoir) do not care whether the q calories went to a system changing reversibly or irreversibly. As long as the heat is extracted at constant temperature, the heat reservoir has changed in a reversible manner. Hence $-q/T$ *is* the entropy increase in the surroundings during the system's irreversible step.

We see that in an irreversible process, q/T is *not* the entropy change of the system, but it *is* the entropy change of the surroundings. This is not a contradiction, though at first glance it may seem so. Within the universe, a part is changing irreversibly (the system), and the rest is changing reversibly (the surroundings). That this is a correct analysis can be most clearly seen for the limiting irreversible process: the expansion of the gas into a vacuum. For this change, "process 0," no work at all is performed and $q = 0$. Yet the initial and final states of the system are the same as for any of the processes pictured in Figure 4-6. For *all* these expansions,

$$\begin{aligned}\Delta S &= S_2 - S_1\\ &= \frac{q_{\text{rev}}}{T}\\ &= \frac{1.40PV}{T}\end{aligned}$$

However, consider the surroundings for this case. For process 0, no heat at all is extracted from the heat reservoir and no work is delivered to the surroundings. There is no doubt at all that the entropy change of the surroundings is zero: since *process 0 causes no change whatsoever in the surroundings, the entropy (randomness) of the surroundings cannot have changed.* We see that ΔS_{surr} must be correctly given by $\Delta S_{\text{surr}} = q/T = 0$.

With this reassurance that $\Delta S_{\text{surr}} = q/T$, let us recapitulate our entropy ledger for any one of the irreversible expansions.

For the system:

heat absorbed $= q$

$$\text{entropy change} = \frac{q_{\text{rev}}}{T} \qquad (4\text{-}52)$$

For the surroundings:

$$\text{heat absorbed} = -q \tag{4-56}$$

$$\text{entropy change} = -\frac{q}{T} \tag{4-57}$$

For the universe:

$$\begin{aligned}\Delta S_{\text{universe}} &= \Delta S_{\text{system}} + \Delta S_{\text{surroundings}} \\ &= \left(\frac{q_{\text{rev}}}{T}\right) + \left(-\frac{q}{T}\right) \\ &= \frac{q_{\text{rev}} - q}{T}\end{aligned} \tag{4-58}$$

This expression is reminiscent of what we called the "extent of irreversibility," expression (4-50). We see that $(q_{\text{rev}} - q)$ is a measure of tendency for spontaneous change in a constant temperature process because, when divided by T, it becomes the Second Law criterion for spontaneous change (4-49). If $(q_{\text{rev}} - q)$ is positive, the entropy of the universe increases and the change can take place spontaneously.

For a spontaneous change, at constant T	$q_{\text{rev}} - q > 0$	*(4-59)*

(b) CONSTANT PRESSURE–CONSTANT TEMPERATURE PROCESSES

The result (4-59) is particularly valuable when applied to a constant pressure–constant temperature process. This type of change is one of the most important to a chemist because so many chemical changes take place in open beakers. The pressure then is automatically constant at one atmosphere, and to keep the temperature constant is very easy. For these interesting processes, the expansion work is extremely easy to calculate. At constant pressure,

$$\text{expansion work} = P(V_2 - V_1) = P\Delta V \tag{4-60}$$

Further, if the reaction proceeds merely in a beaker (not in an electrochemical cell), expansion work is the only kind of work performed. As we have shown in Chapter Two (*see* Section 2-3(d)),

$$\Delta H = q_P \tag{2-27}$$

Substituting q_p into our measure of reaction tendency $(q_{rev} - q) > 0$, we obtain:

$$q_{rev} - q > 0 \quad \text{for a spontaneous change at constant } T$$
$$q_{rev} - q_P > 0 \quad \text{for a spontaneous change at constant } T \text{ and } P \qquad (4\text{-}61)$$

Also, by substituting (2-27) and (4-47) into (4-61), we obtain

$$T\Delta S - \Delta H > 0 \quad \text{for a spontaneous change at constant } T \text{ and } P$$

or,

$$T(S_2 - S_1) - (H_2 - H_1) > 0$$

or,

$$(TS_2 - H_2) - (TS_1 - H_1) > 0 \qquad (4\text{-}62)$$

Since both S and H are "functions of state," $(TS - H)$ must be also. This quantity, or rather its negative, is called the *free energy* and is designated these days by the symbol G.

$$\textbf{free energy} = G = H - TS \qquad (4\text{-}63)$$

Substituting (4-63) into (4-62), we obtain

$$(-G_2) - (-G_1) > 0$$

or,

$$G_2 - G_1 < 0.$$

For a spontaneous change at constant T and P $\quad \Delta G < 0$ $\qquad$ *(4-64)*

We see that a spontaneous change (constant T,P) is accompanied by a decrease in G. The connection to $(q_{rev} - q)$ shows that when ΔG is negative, the entropy of the universe increases. *The free energy is a quantitative measure of reaction tendency.* Its advantage is that $\Delta G = \Delta H - T\Delta S$ gives us a criterion for spontaneous change in which we need look only at properties of our system. This is sufficient to take account of the rest of the universe.

Free energy change has one more important property. Remember that we define ΔG in such a way that, for a change at constant pressure and temperature (with expansion work only), we would have

$$\Delta G = q_P - q_{rev} \tag{4-65}$$

If this relation is applied to a change that occurs *at equilibrium*, while maintaining constant pressure and temperature, then q_P becomes equal to q_{rev}.

For an equilibrium process	$\Delta G = q_{rev} - q_{rev} = 0$

(4-66)

This would be the case, for example, when a liquid vaporizes at a constant pressure equal to its vapor pressure at the temperature of the thermostat. It would be applicable to a chemical reaction occurring hypothetically in an equilibrium mixture. Most important of all, it is a crossover point between conditions in which a reaction will take place spontaneously as written and conditions in which the reaction will take place spontaneously in the reverse direction.

If ΔG is negative, $\Delta G < 0$	change will occur spontaneously

(4-67)

If $\Delta G = 0$	equilibrium exists

(4-68)

If ΔG is positive, $\Delta G > 0$	change cannot occur spontaneously as written—reverse change will occur spontaneously

(4-69)

(c) THE MEANING OF ΔG

We have achieved an important advance. The Second Law of Thermodynamics tells us unequivocally that *all* processes proceed spontaneously in the direction that raises the entropy (randomness) of the universe. However, the universe is a big place, and it's hard to keep track of its parts. Now, with the aid of free energy, we can look at only properties of the system. For changes that occur at constant temperature and pressure, we can be sure that the entropy of the universe goes up if the free energy of the system goes down.

The free energy criterion is the one that chemists most commonly use, and it is often posed as a measure of competition between an energy effect and a randomness effect. In our new expression for reaction tendency,

$$\Delta G = \Delta H - T\Delta S \tag{4-70}$$
$$\Delta G < 0 \quad \text{indicates a spontaneous reaction} \tag{4-71}$$

Notice that ΔG is made more negative (spontaneous direction) by both a negative value of ΔH (exothermic reaction) and it is made more negative by a positive value of ΔS (more randomness in the system). Hence, the free energy criterion seems to raise a conflict of desires between a reaction's tendency to proceed in the exothermic direction (that was Berthelot's proposal) and the reaction's tendency to proceed to the most random state. This is a workable basis upon which to predict chemistry, and it will be done in this fashion for some time to come. Nevertheless, it is desirable to remember that the Second Law makes no reference whatsoever to energy effects—*spontaneity is governed solely by tendency to randomness.*

Let us be reminded again of the way in which $\Delta G = \Delta H - T\Delta S$ implements the randomness principle.

First, let us divide each term in (4-70) by $-T$:

$$\left(-\frac{\Delta G}{T}\right) = -\frac{\Delta H}{T} + \Delta S \tag{4-72}$$

Now we see a sum of quantities, all with the dimensions of entropy (calories/mole degK). On the left-hand side we have the negative of $\Delta G/T$. Remembering that ΔG must be negative for a spontaneous change, we see that $(-\Delta G/T)$ must be positive. On the right-hand side we see two contributions; the second being clearly the entropy of the system. Its contribution is obvious—if the entropy of the system goes up (counting both the system's positional and motional randomness), the entropy of the universe tends to go up. A positive ΔS tends to urge the reaction to proceed spontaneously. All we need look into now, is the entropy change of the rest of the universe. That must be embodied in the $-\Delta H/T$ term. How does the exothermic heat of a reaction reflect the entropy of the rest of the universe? It does so for the special case of a constant temperature–constant pressure process, because, if temperature is to remain constant, the exothermic enthalpy must be taken up as heat in the surroundings. Converting an amount of potential energy ΔH into heat creates motional randomness in the surroundings, in amount $(-\Delta H/T)$. Thus $(-\Delta G/T)$ equals the entropy change of the universe (for constant T,P processes), and it is a direct implementation of the Second Law.

$\left(-\frac{\Delta G}{T}\right)$	=	$\left(-\frac{\Delta H}{T}\right)$	+	ΔS	(4-73)
entropy rise for the universe	=	entropy rise due to motional randomness in the surroundings	+	entropy rise due to all kinds of randomness in the system	

If $(-\Delta G/T)$ is positive, the entropy of the universe increases, the universe becomes more random, and the change will occur spontaneously. In order for $(-\Delta G/T)$ to be positive, ΔG must be a negative quantity. Hence, $\Delta G < 0$ for the system is a criterion for spontaneous change.

(d) REACTION IN A BULB

Our interest in constant pressure processes is justified by the fact that in practice many chemical processes are carried out in open containers, hence, under constant pressure conditions. On the other hand, many chemical changes *must* be carried out in closed containers for control of a variety of conditions—specially low or specially high pressure may be needed; exposure to air may cause interfering reactions; physical handling of the substances may be facilitated. Chemistry carried out in a closed bulb occurs, of course, at constant volume and not necessarily at constant pressure.

To see the significance of this difference, consider a constant temperature–constant volume process. The First Law expression becomes especially simple (if no electrical work is performed), because expansion work requires a change in volume. For a reaction in a bulb, work is zero.

$$\Delta E = q - w = q_V \qquad (2\text{-}22)$$

With this start, let's return to our "extent of irreversibility;" the implementation of the Second Law.

Substituting (4-47) and (2-22) into (4-50), we get

$$\text{extent of irreversibility} = q_{\text{rev}} - q = T\Delta S - q_{\text{v}} = T\Delta S - \Delta E \qquad (4\text{-}74)$$

Remembering (2-65), which for constant volume becomes $\Delta H = \Delta E + V\Delta P$, (4-74) can be written as

$$q_{\text{rev}} - q = T\Delta S - (\Delta H - V\Delta P) = (T\Delta S - \Delta H) + V\Delta P \qquad (4\text{-}75)$$

The first part of (4-75) is just the definition of $-\Delta G$ from (4-70). Hence, our spontaneity condition that $q_{\text{rev}} - q$ be greater than zero can be expressed, for a constant volume change, as follows.

$q_{rev} - q = [(-\Delta G) + V\Delta P]$ will be positive

or, reversing the signs

For a spontaneous process at constant volume	$\Delta G - V\Delta P$ will be negative $\Delta G - V\Delta P < 0$	*(4-76)*

Expression (4-76) is the general criterion for a spontaneous change at constant volume. It becomes identical to the constant pressure criterion, (4-71), for those cases in which $\Delta P = 0$, despite the fact that the reaction occurred in a closed bulb. There are two common situations in which this happens. For a change that occurs completely among condensed phases (liquids and solids), the density change is almost always negligible. Then the pressure does not change even though no attempt is made to keep it constant. For such situations, including the host of reactions that take place in solution, for all practical purposes $\Delta P = 0$ and $\Delta G - V\Delta P = \Delta G$.

If gases are involved, the value of $V\Delta P$ depends upon the difference between the number of moles of gas consumed, $n_{\text{reactants}}$, and the number of moles of gas produced, n_{products}. As shown in Chapter Two, expression (2-68),

$$V\Delta P = (n_{\text{products}} - n_{\text{reactants}})RT$$

$$V\Delta P = \Delta nRT \qquad (2\text{-}68)$$

For many reactions involving gases, $\Delta n = 0$; that is the number of moles consumed happens to equal the number of moles produced. The reaction $H_2(g) + I_2(g) = 2HI(g)$ is such a case. Then, $V\Delta P = \Delta nRT = 0$ and, once again, $\Delta G - V\Delta P = \Delta G$.

In summary, for a constant volume process, (4-76) is the spontaneity criterion, although, for many chemical processes, $\Delta P = 0$ and (4-76) becomes identical to (4-71).

4-5 Some examples: Enthalpy, entropy, and free energy

Having concluded that free energy ΔG is a measure of reaction tendency, it behooves us to look at a few examples. Free energy is a composite of enthalpy and entropy, and, for reactions at constant temperature and pressure, these are both simply related to thermal effects. The enthalpy change can be directly measured in a calorimeter, as described in Section 2-4, and it is the heat at constant pressure, q_p, that is desired,

$$\Delta H = q_P \qquad (2\text{-}27)$$

The entropy change is not so simply measured in a calorimeter, because it is the *reversible* heat that is needed, and reactions that proceed spontaneously in a calorimeter do not proceed reversibly. Nevertheless, there are several good ways to measure ΔS and some of them will be mentioned in the next chapter. For our purposes here, we'll merely observe that there are lengthy tables of measured entropies available for our use. If, then, we can find a tabulation of measured enthalpies and entropies, we should be able to make quick predictions concerning reaction tendency. Here are some examples.

(a) THE MELTING OF ICE: $H_2O(s) \rightarrow H_2O(\ell)$

At 0°C (273°K), the heat of melting (fusion) of ice has been measured innumerable times and is known with high accuracy. This constant pressure heat equals the enthalpy of melting:

$$\Delta H_{273} = q_{\text{melting}} = +1436 \text{ calories/mole}$$

Equally well known is the entropy of melting:

$$\Delta S_{273} = +5.257 \text{ calories/mole degK}$$

According to equation (4-70), the free energy change associated with melting can now be calculated:

$$\begin{aligned} \Delta G_{273} &= \Delta H - T\Delta S \\ &= (+1436) - (273.16)(5.257) \\ &= +1436 - 1436 \\ &= 0 \end{aligned}$$

Since the free energy change is zero, we see that ice at 273°K and water at 273°K are at equilibrium.

Now let us consider the same change, but at 25°C (298°K). Here again, experimental measurements show that

$$\begin{aligned} \Delta H_{298} &= +1669 \text{ calories/mole} \\ \Delta S_{298} &= +6.076 \text{ calories/mole degK} \end{aligned}$$

We can calculate the free-energy change:

$$\begin{aligned} \Delta G_{298} &= \Delta H_{298} - T\Delta S_{298} \\ &= +1669 - (298.16)(6.076) \\ &= -142 \text{ calories/mole} \end{aligned}$$

The free energy change is negative—the process will occur spontaneously. The free energy change tells us what we already knew—ice melts at 25°C.

It is instructive to examine the signs and magnitudes of the contributions to ΔG. The value of ΔH is positive—heat is absorbed, so the ice, as it melts, is "rolling uphill" as far as potential energy is concerned. The surroundings must surrender heat, so the entropy of the surroundings opposes reaction. Nevertheless, the entropy of the system more than overcomes this effect at 25°C, and impels the melting process. The positive sign of ΔS implies that the melting process tends to a state of higher probability. Our knowledge of the regular nature of a solid and the random disorder of a liquid makes this intuitively reasonable. This situation *always* exists for a melting process; the regular crystal lattice has lower potential energy, so ΔH is positive, but the positional randomness of the liquid, expressed in a positive ΔS, favors melting.

Now we can look at the melting point in a new light. As the temperature rises, both ΔH and ΔS change quite slowly. There is, however, the additional rapid dependence upon temperature in the entropy term $T\Delta S$. Because of this multiplication by T the $T\Delta S$ term becomes unimportant at low temperatures. The ΔH term then dominates ΔG, and the solid is the stable form. As T rises, there will be a temperature at which $T\Delta S$ exactly equals the potential energy contribution to ΔG; then $\Delta G = 0$. At this temperature, both solid and liquid are stable and they can coexist. This is the melting point. At still higher temperatures, the $T\Delta S$ term dominates, ΔG is negative, and the liquid becomes the thermodynamically stable form.

(b) VAPORIZATION OF LIQUID WATER

Table 4-4 shows the values of ΔH and ΔS for the vaporization of water, both at 100°C (373°K) and 25°C (298°K). In each case, we must specify the pressure of the water vapor, since the entropy of a gas depends significantly upon its pressure. As the pressure of a gas rises, its intrinsic randomness decreases. Putting this in more intuitive terms, as the pressure rises, a gas becomes more and more like a liquid, and the difference in randomness between a gas and its liquid form tends to decrease.

Table 4-4 Thermodynamics of the Vaporization of Water

$$H_2O(liquid) \xrightarrow[T,P]{} H_2O(gas)$$

Temperature	100°C, 373°K	25°C, 298°K	25°C, 298°K
H_2O vapor pressure (torr)	760	760	23.8
ΔH (cal/mole)	+9,721	+10,489	+10,489
ΔS (cal/mole degK)	+26.05	+28.26	+35.18
$\Delta G = \Delta H - T\Delta S$ (cal/mole)	0	+2,063	0

Consulting the last line of Table 4-4, we see that at 100°C the vaporization of liquid water to gas at a pressure of 760 torr is an equilibrium process; $\Delta G = 0$. Liquid water and water vapor at one atmosphere can coexist. However, the same change carried out at 25°C, also at one atmosphere pressure, has a large, *positive* ΔG. A spontaneous change will occur when ΔG is negative; so to obtain a spontaneous change, we must consider the reverse change!

$$\text{if} \quad H_2O(\ell) \xrightarrow{298°K} \underset{760\ \text{torr}}{H_2O(g)} \quad \Delta G = +2{,}063$$

$$\text{then} \quad \underset{760\ \text{torr}}{H_2O(g)} \xrightarrow{298°K} H_2O(\ell) \quad \Delta G = -2{,}063$$

We see that the *free energy predicts the spontaneous condensation of water vapor to the liquid.* The last column of Table 4-4 indicates how long this condensation process will continue to occur spontaneously. As condensation proceeds, the vapor pressure drops, and the entropy difference between liquid and gas increases. Finally, at 23.8 torr, the entropy change rises to the point at which $T\Delta S$ becomes as large as ΔH. Now $\Delta G = 0$, and hence, equilibrium is restored. The pressure 23.8 torr is the equilibrium vapor pressure of water at 25°C.

Let us consider again the magnitudes of ΔH and ΔS. Notice that the heat effect is more than five times that of the melting process. This is because the melting process leaves the molecules close together, at distances fixed by attractive forces—it costs only the extra potential energy of the long-range crystal regularity. Vaporization, however, requires that the molecules be pulled away from each other, a more expensive process.

The randomness effects in vaporization are also larger. The gas phase, characteristically, gives the molecules more positional randomness than the liquid, because of the molecular freedom of movement in a gas. This factor is, however, sensitively dependent upon the pressure, and at any temperature, there is a particular pressure at which $\Delta H = T\Delta S$, so that $\Delta G = 0$. This criterion fixes the equilibrium vapor pressure.

(c) SOME CHEMICAL CHANGES INVOLVING SOLIDS

Here are two familiar chemical reactions involving both gases and solids.

$$CaCO_3(\text{solid}) \xrightarrow{298°K} CaO(\text{solid}) + \underset{1\ \text{atm}}{CO_2(\text{gas})} \qquad (4\text{-}77)$$

$$C(\text{graphite}) + \underset{1\ \text{atm}}{O_2(\text{gas})} \xrightarrow{298°K} \underset{1\ \text{atm}}{CO_2(\text{gas})} \qquad (4\text{-}78)$$

Equation (4-77) is concerned with the stability of limestone (calcium carbonate), a stable solid found in nature everywhere. Equation (4-78) relates to the oxidation of carbon at 25°C, obviously a reaction connected to the flammability of carbon and the contrasting stability of the graphite writing core of an ordinary pencil. Table 4-5 shows the thermodynamic factors in each reaction.

Table 4-5 Thermodynamic Properties of Two Chemical Reactions Involving Solids ($T = 298°K$)

$CaCO_3(s) \rightarrow CaO(s) + CO_2(1\ atm,\ g)$		$C(s) + O_2(1\ atm,\ g) \rightarrow CO_2(1\ atm,\ g)$
ΔH(kcal/mole)	+42.55	−94.05
ΔS(kcal/mole degK)	+38.4	+ 0.73
ΔG(kcal/mole)	+31.10	−94.27

The first reaction has a positive ΔG, and the significance of this is in accord with experience. A change with positive ΔG proceeds spontaneously in the *reverse* direction. In other words, calcium oxide reacts with carbon dioxide at one atmosphere pressure to give calcium carbonate. This occurs in spite of the large positive ΔS that accompanies the formation of a gaseous product, with its high positional randomness. Of course, the effect of this entropy factor will increase markedly if the temperature is raised. There is some temperature above which limestone becomes unstable with respect to reaction (4-77). Experiments, including thermodynamic measurements, show that this temperature is near 850°C. Calculations of this kind are helpful in designing a furnace to convert limestone ($CaCO_3$) into lime (CaO) for use in plaster. In fact, lime is manufactured in this way, by heating limestone to temperatures in excess of 800°C (to the tune of millions of tons per year in the U.S. alone).

The second reaction has a startling result. Since ΔG is negative, our thermodynamic reasoning indicates a tendency for reaction to occur. Graphite has lots of tendency to react (to "burn") with the oxygen in the atmosphere. For example, if the graphite in your pencil understood its thermodynamics, it would react promptly and release a large amount of heat.

Here we see one of the most important distinctions that a chemist must keep in mind. There are two facets of chemical change that interest him, that must be understood, and that can be manipulated to advantage. First, there is the direction of spontaneous change, and secondly, the rate of the change. Thermodynamic arguments are essentially equilibrium arguments; they give us a signpost toward equilibrium. They tell nothing, however, of the mountains that will be encountered on the way. These interesting mountains (free energy mountains) determine the rate. Obviously it is to our advantage to know that graphite is "ready" to react with air to release energy generously, if we also know how to speed up the reaction when the energy is needed.

There is another valuable aspect to the graphite oxidation. Observe, in Table 4-5, the very small entropy contribution to the reaction tendency, particularly compared with that of the limestone decomposition. This small value of ΔS reflects the fact that one mole of gas is consumed as one mole is produced. The randomness associated with a gaseous product is almost exactly balanced by the randomness of an equal number of moles of a gaseous reactant.

(d) MORE CHEMICAL CHANGES INVOLVING GASES ONLY

Table 4-6 lists four gas-phase reactions that provide some interesting contrasts. First, note that the four reactions have been selected for their similarity. In each, two moles of a triatomic molecule react to produce three moles of diatomic molecules. The production of three gaseous molecules of two types from only two gaseous molecules of a single type should increase the randomness considerably. The values of ΔS should all be large and positive—as they are.

*Table 4-6 Thermodynamic Properties of Some Simple Reactions Involving Gases**

Reaction (298°K)	ΔH (kcal/mole)	ΔS (cal/mole degK)	ΔG (kcal/mole)	
$2CO_2(g) \rightarrow 2CO(g) + O_2(g)$	+135.3	+41.5	+122.9	*(4-79)*
$2N_2O(g) \rightarrow 2N_2(g) + O_2(g)$	−39.0	+35.4	−49.5	*(4-80)*
$2NO_2(g) \rightarrow 2NO(g) + O_2(g)$	+27.0	+34.7	+16.7	*(4-81)*
$2NO_2(g) \rightarrow 2O_2(g) + N_2(g)$	−16.2	+28.8	−24.8	*(4-82)*

*All pressures, one atmosphere.

The ΔH column shows dramatic variation, however, and these differences are influential in determining the sign and magnitude of ΔG, hence the reaction tendency. The large positive ΔH for decomposition of carbon dioxide causes the free energy change to be positive, so the reaction proceeds spontaneously to the left if all constituents are at 1-atmosphere pressure. The opposite is true for N_2O. Here the negative ΔH causes ΔG to be negative, and we conclude that N_2O is unstable with respect to decomposition. Its availability as a common laboratory chemical depends upon kinetic effects. Under suitable conditions or in the presence of a catalyst, the spontaneous tendency for N_2O to decompose can be used to advantage.

The important conclusion is that the reaction heat (the enthalpy change) is of crucial importance to us. This macroscopic effect, the difference between reactions (4-79) and (4-80), must be understood on the microscopic level in order to understand and predict chemical reactions. It is because of differences such as in the ΔH values for CO_2 and N_2O in Table 4-6, that chemists are so concerned about the chemical bonding in molecules.

The last pair of reactions in Table 4-6, (4-81) and (4-82), also contains an interesting lesson. Here we see a molecule that might decompose in two ways: either to NO plus O_2, or to N_2 plus O_2. Thermodynamic measurements show that NO_2 will not spontaneously form NO plus O_2 at one atmosphere pressure for each constituent—no, the opposite reaction might occur. On the other hand, NO_2 *is* unstable with respect to formation of N_2 and O_2; the free energy change for reaction (4-82) is negative. Again the explanations of these differences must be sought in the bonding of the molecules. For the moment we can observe the potential value of thermodynamic arguments in terms of a practical and terribly pressing human problem. One of our growing concerns is the air pollution in cities. Gasoline-powered automobiles daily spew tons of nitric oxide into the atmosphere we breathe. The reactions of Table 4-6 might contain a solution. According to Table 4-6, nitric oxide can spontaneously react with oxygen to form NO_2. In turn, NO_2 is unstable with respect to decomposition into N_2 and O_2, both of which are quite acceptable atmospheric constituents. Thus, our attention is focussed on finding conditions that would accelerate reaction (4-82). With disconcerting confidence, we can predict that smog, like other threats to human existence, will ultimately yield in the face of our understanding of thermodynamics, reaction kinetics, and chemical bonding.

five
free energy and equilibrium

At constant pressure and temperature, spontaneous processes seek the lowest possible free energy. In such cases, lowering the free energy of a system corresponds to raising the entropy of the universe. Hence, we can say that on a free energy landscape, chemical changes tend to "roll downhill." If a postulated change has positive ΔG, the reaction is heading uphill, and we can be sure the reaction will in fact go in the reverse direction (reaction rate being willing). When $\Delta G = 0$, there is no tendency to go in either direction. At the bottom of the free energy valley, equilibrium exists.

5-1 Downhill is the only way to go

This "rolling downhill" principle is useful in clarifying equilibrium, as well as the approach to equilibrium. Let's explore it in detail.

(a) FREE ENERGY, AN EXTENSIVE PROPERTY

A property that is proportional to the amount of substance is called an "extensive" property (it depends upon the "extent" of the system). Mass is an obvious example. One mole of liquid ethanol, C_2H_5OH, weighs 46 grams; two moles weigh $2 \cdot 46 =$ 92 grams; ten moles weigh 460 grams. (A familiar example of a property that is *not* extensive is the density. The density of one mole of ethanol is 0.789 g/ml; the same as the density of two or ten moles.) Free energy is an extensive property. It is the sum of an energy factor and a probability factor. The energy of a system is clearly determined by the extent of the system: the kinetic energy of two moles of gas at a given temperature is double that of one mole of the gas, since there are more particles dashing about. The potential energy in a kilogram of sodium chloride is a thousand times that of a gram.

The probability factor, too, is extensive. Entropy is defined as the logarithm of probability. Independent probabilities combine in a multiplicative way, and the logarithm of a product equals the sum of the logarithms of the factors (see Section 4-3). Two moles of a gas can be considered as two systems each composed of one mole. Hence, the entropy of two moles is double that of one mole. If both energy and entropy are extensive, their sum must also be extensive, so free energy is an extensive property.

This means we can calculate the free energy of a gas mixture as the sum of the free energies of its constituents, by taking account of the number of moles of each constituent. If, for example, we have three constituents, A, B, and C, we can express the total free energy as

$$G = n_A G_A + n_B G_B + n_C G_C \qquad (5\text{-}1)$$

The quantity G_A is the free energy per mole of substance A, and n_A is the number of moles of A. This expression permits us to follow the free energy as a reaction proceeds.

(b) FREE ENERGY AND THE FORMATION OF HYDROGEN IODIDE

Application of expression (5-1) to a particular reaction is very informative. A simple example is the now familiar reaction between hydrogen and iodine to form hydrogen iodide. Let's again consider this reaction. Beginning with one mole of H_2 at a pressure of 1 torr mixed with one mole of gaseous I_2 also at 1 torr pressure, let's suppose the reaction could occur with complete conversion to two moles of hydrogen iodide at 2 torr pressure. We'll consider the reaction at 700°K, a temperature at which the reaction rate is found to be rapid.

At the start, expression (5-1) will be simple because only H_2 and I_2 are present, one mole of each. At the end of the reaction there will be only HI present, two moles of it. At intermediate points, there will be something of everything present, H_2, I_2, and HI. The amount of HI depends on the amount of H_2 and I_2 lost through the reaction. If x moles of H_2 have reacted, there will be $2x$ moles of HI formed and $(1 - x)$ moles each of H_2 and I_2 left.

$$G = n_{H_2} G_{H_2} + n_{I_2} G_{I_2} + n_{HI} G_{HI} .$$

Initially,

$$G_a = 1 \cdot G_{H_2} + 1 \cdot G_{I_2} + 0 \cdot G_{HI} = G_{H_2} + G_{I_2} \qquad (5\text{-}2)$$

finally,

$$G_z = 0 \cdot G_{H_2} + 0 \cdot G_{I_2} + 2 \cdot G_{HI} = 2 \cdot G_{HI} \qquad (5\text{-}3)$$

and in between, when x moles of H_2 have reacted,

$$G_x = (1 - x) \cdot G_{H_2} + (1 - x) \cdot G_{I_2} + (2x) \cdot G_{HI} \qquad (5\text{-}4)$$

If G_a, G_z and some intermediate values of G_x were evaluated, they would show how the free energy changes during the reaction. To do this, it is helpful to consider the two components of free energy, $H - TS$, separately. Each of the expressions (5-2),

(5-3), and (5-4) becomes more informative and useful when written as the sum of an enthalpy and an entropy term.

$$G_a = G_{H_2} + G_{I_2} = (H_{H_2} - TS_{H_2}) + (H_{I_2} - TS_{I_2})$$

or

$$G_a = (H_{H_2} + H_{I_2}) - T(S_{H_2} + S_{I_2}) \tag{5-5}$$

Also,

$$G_z = 2H_{HI} - 2TS_{HI} \tag{5-6}$$

and

$$\begin{aligned} G_x = {} & [(1-x)H_{H_2} + (1-x)H_{I_2} + 2xH_{HI}] \\ & - T[(1-x)S_{H_2} + (1-x)S_{I_2} + 2xS_{HI}] \end{aligned} \tag{5-7}$$

(c) THE ENTHALPY PART

Let's wrestle first with the enthalpy parts of expressions (5-5), (5-6) and (5-7). What is the magnitude of $(H_{H_2} + H_{I_2})$ in (5-5)? The enthalpy of one mole of iodine gas at 700°K is a composite of many energy contributions. First, the molecules have a variety of translational energies that average out to $(3/2)RT$ ergs per mole. The molecules also rotate and vibrate; so there are two more energy contributions. In addition to these "energy of motion" quantities, the iodine molecule possesses "energy of position;" that is, each molecule consists of two atoms spaced at the equilibrium distance at which their energy is a minimum. Probing further, the electrons in the atoms have energy of motion and energy of position. And finally, to compound the problem to the utmost, the nuclei have enormous, though little understood, binding energies holding each one together.

So just to express the enthalpy of a mole of iodine relative to its primordial constituents at 0°K, we set off on quite an odyssey. The situation can be likened to a building contractor in Denver, Colorado, who wonders whether his tract site is above or below the water reservoir on the other side of the valley (see Fig. 5-1). He could try to determine the altitude of the reservoir relative to sea level out in California. On the other hand, he might decide he hasn't time to triangulate his way over the Rockies and all the way to Laguna Beach. It will suffice to pick some convenient local reference point and forget about sea level. In fact, the reservoir itself might as well be taken as the reference altitude.

We will follow the home-builder's lead. For the present purpose, let's calculate all enthalpies relative to the elements gaseous H_2 and gaseous I_2, each at 1 torr pressure and 700°K. This will be our zero of enthalpy.

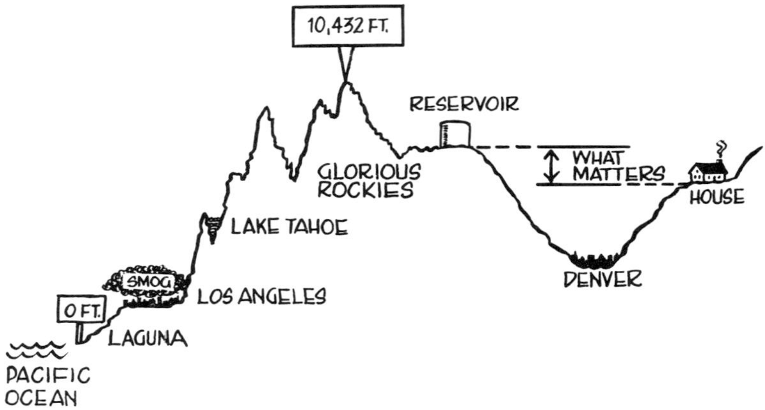

Figure 5-1 It's all relative, anyway.

Now the enthalpy parts of (5-5), (5-6), and (5-7) are quite easy to determine.

$$H_a = H_{H_2} + H_{I_2} = 0 + 0 = 0 \qquad (5\text{-}8)$$

and

$$H_z = 2H_{HI} \qquad (5\text{-}9)$$

The magnitude of (5-9) depends upon the enthalpy of two moles of gaseous hydrogen iodide with reference to the elements H_2 and I_2. Experimentally, this is what is measured with a calorimeter in which known amounts of H_2 and I_2 react at constant temperature and constant pressure to form HI. For example, half a mole of H_2 could be mixed with half a mole of I_2, each at a pressure of 1 torr. Then when reaction occurs, one mole of HI is formed at a pressure of 2 torr. The total pressure will be constant at 2 torr throughout the reaction, and in a thermostatted calorimeter, the temperature can be held constant. The reaction heat, under these conditions, has been measured.

$$\tfrac{1}{2}H_2 \text{ (1 torr)} + \tfrac{1}{2}I_2 \text{ (1 torr)} \rightarrow HI \text{ (2 torr)} \qquad (5\text{-}10)$$
$$\Delta H = -1.59 \text{ kcal}$$

The reaction is exothermic, so the potential energy of HI is below that of its constituent elements. Substituting (5-10) into (5-9),

$$H_z = 2(-1.59) = -3.18 \text{ kcal} \qquad (5\text{-}11)$$

At any intermediate point, we can easily evaluate H_x:

$$H_x = (1 - x)H_{H_2} + (1 - x)H_{I_2} + 2xH_{HI}$$

For example, when 75 percent of the H_2 has reacted, $x = 0.75$, so

$$\begin{aligned} H_{0.75} &= (1 - 0.75) \cdot H_{H_2} + (1 - 0.75) \cdot H_{I_2} + 2 \cdot (0.75) \cdot H_{HI} \\ &= (0.25) \cdot (0) + (0.25) \cdot (0) + (1.50) \cdot (-1.59) \\ &= -2.38 \text{ kcal} \end{aligned} \qquad (5\text{-}12)$$

Figure 5-2 shows a graph of the change of total enthalpy as reaction proceeds. It is simply a straight-line relationship. It tends to cause ΔG to go "downhill" as H_2 and I_2 are consumed and HI is produced. We see that the enthalpy term favors the formation of HI, and as far as enthalpy is concerned, the reaction should proceed to completion, consuming all of the H_2 and I_2.

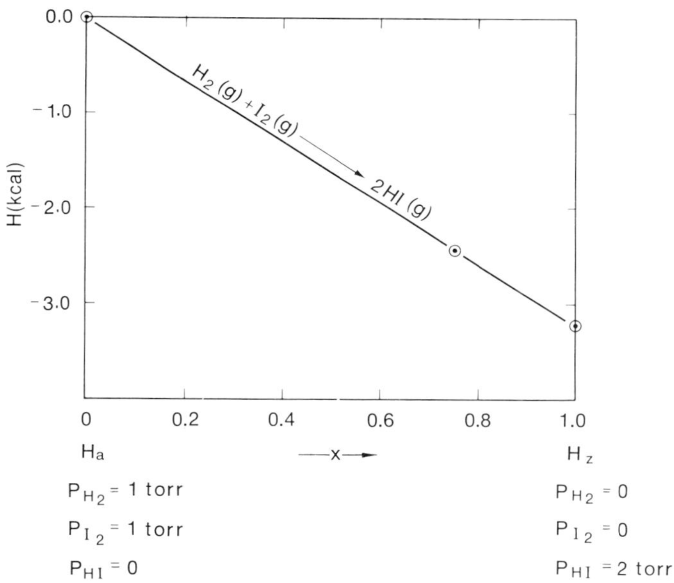

Figure 5-2 Total enthalpy during the reaction

$$H_2 \text{ (1 torr)} + I_2 \text{ (1 torr)} \xrightarrow{700^\circ K} 2HI \text{ (2 torr)}$$

(d) DON'T FORGET ENTROPY

We'll take the same approach to entropy as we did to enthalpy. Rather than place things on an absolute (sea-level) scale, we'll again take our starting conditions of H_2 at 1 torr pressure and I_2 at 1 torr pressure, as our reference point. The entropy parts of (5-5), (5-6), and (5-7) can now be determined.

$$S_a = S_{H_2} + S_{I_2} = 0 + 0 = 0 \quad (5\text{-}13)$$

and

$$S_z = 2S_{HI} \quad (5\text{-}14)$$

Again we'll appeal to experimental measurements to give us the entropy of HI at the end of the reaction. The experimental result is 0.440 cal/degK per mole of HI at 2 torr pressure.

$$S_z = 0 \cdot S_{H_2} + 0 \cdot S_{I_2} + 2 \cdot S_{HI} = 2 \cdot (0.440)$$
$$= 0.880 \text{ cal/degK} \quad (5\text{-}15)$$

Unfortunately, to calculate the entropy at intermediate stages of the reaction is more complicated than for enthalpy. This is because the entropy per mole of gas depends upon its pressure, whereas the energy and enthalpy content of a mole of perfect gas are independent of pressure at constant temperature. The entropy (randomness) of a gas increases as the volume available to each molecule increases.

The manner in which this effect contributes to the value of S can be illustrated by one particular case, in which, say, three-quarters of the reaction has occurred. At this point in the reaction, $\frac{3}{4}$ mole of H_2 has reacted with $\frac{3}{4}$ mole of I_2. This leaves $\frac{1}{4}$ mole of each that has not reacted, but at one-quarter the initial pressure. The conditions in the reaction vessel are as follows:

$$n_{H_2} = 0.25 \text{ moles} \qquad p_{H_2} = 0.25 \text{ torr}$$
$$n_{I_2} = 0.25 \text{ moles} \qquad p_{I_2} = 0.25 \text{ torr}$$
$$n_{HI} = 1.5 \text{ moles} \qquad p_{HI} = 1.5 \text{ torr}$$

The entropy of the 0.25 moles of H_2 at 0.25 torr is different from the entropy of 0.25 moles of H_2 at 1.0 torr (which we took to be zero). The change is not difficult to evaluate since we are concerned with a constant temperature expansion of a perfect gas. The change we need to consider is

$$\begin{bmatrix} 0.25 \text{ moles } H_2 \\ P_1 = 1.0 \text{ torr} \\ T = 700°\text{K} \end{bmatrix} \xrightarrow[\Delta S]{\Delta E} \begin{bmatrix} 0.25 \text{ moles } H_2 \\ P_2 = 0.25 \text{ torr} \\ T = 700°\text{K} \end{bmatrix} \quad (5\text{-}16)$$

We can immediately conclude that $\Delta E = 0$, since H_2 is a perfect gas, and at constant temperature the energy of a perfect gas does not change. This result, combined with the First Law of Thermodynamics, permits us to calculate the heat absorbed.

$$\Delta E = q - w$$

but

$$0 = q - w$$

$$q = w \quad (5\text{-}17)$$

Now suppose the expansion is carried out reversibly. This gives us the maximum possible work and the reversible heat, as was shown in Section 4-1.

For a reversible expansion,

$$q = q_{\text{rev}}$$

and

$$w = w_{\max} = nRT \log_e \frac{V_2}{V_1}$$

therefore,

$$q_{\text{rev}} = nRT \log_e \frac{V_2}{V_1} \qquad (5\text{-}18)$$

For the change as specified in (5-16), it is more convenient to express (5-18) in terms of pressure using the perfect gas relation $P_1V_1 = P_2V_2$.

$$q_{\text{rev}} = nRT \log_e \frac{P_1}{P_2} \qquad (5\text{-}19)$$

Expression (5-19) is just what we need to calculate ΔS for (5-16), since, for a constant temperature process,

$$\Delta S = \frac{q_{\text{rev}}}{T} \qquad (5\text{-}20)$$

so

$$\Delta S = \frac{nRT \log_e(P_1/P_2)}{T}$$

$$= nR \log_e \frac{P_1}{P_2} \qquad (5\text{-}21)$$

Now we can calculate the entropy of one mole of H_2 at 0.25 torr pressure (relative to 1.0 torr). Inserting $n = 1$, $R = 1.99$ cal/mole degK, $P_1 = 1$ torr, and $P_2 = \frac{1}{4}$ torr,

$$S = (1) \cdot (1.99) \cdot \log_e \frac{1}{\frac{1}{4}} = 1.99 \log_e 4$$

$$= 2.76 \text{ cal/degK per mole } H_2 \qquad (5\text{-}22)$$

The iodine gas has the same entropy rise per mole, since its pressure is also reduced from 1 torr to 0.25 torr. Applying (5-21) to the HI gas, we see that its entropy rises (per mole of HI) by 0.572 cal/degK as its pressure is reduced from 2 torr to 1.5 torr.

Now the entropy for $x = 0.75$ can be calculated

$$
\begin{aligned}
S_{0.75} &= n_{H_2} \cdot (S_{H_2,0.25\text{torr}}) + n_{I_2}(S_{I_2,0.25\text{torr}}) + n_{HI}(S_{HI,1.5\text{torr}}) \\
&= n_{H_2}(S_{H_2,1\text{torr}} + R \log_e 4) \\
&\quad + n_{I_2}\ (S_{I_2,1\text{torr}} + R \log_e 4) \\
&\quad + n_{HI}(S_{HI,2\text{torr}} + R \log_e \tfrac{2}{1.5}) \\
&= (0.25)(0 + 2.76) + (0.25)(0 + 2.76) + (1.5)(0.44 + 0.572) \\
&= 2.90 \text{ cal/degK} \qquad (5\text{-}23)
\end{aligned}
$$

Recapitulating, we took as our reference point $S_a = 0$. Experimentally, we determined that $S_Z = +0.88$ cal/degK. Now we have calculated that when three-quarters of the reaction has occurred, $S_{0.25} = +2.90$ cal/degK. This is higher than either S_a or S_z! We see that a plot of S_x against x will have a more complicated shape than the H_x plot of Figure 5-2. Calculations for different values of x lead to Figure 5-3, a consequence of the fact that randomness increases as gas pressure drops, as expressed in (5-21).

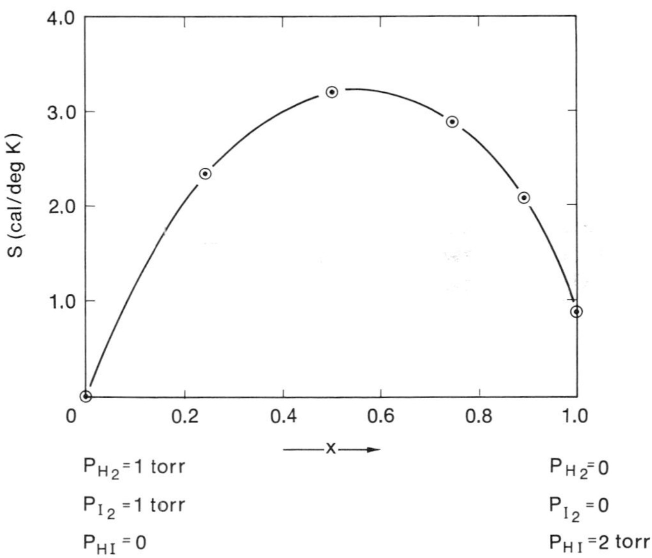

Figure 5-3 Total entropy during the reaction

$$H_2\ (1\ \text{torr}) + I_2\ (1\ \text{torr}) \xrightarrow{700°K} 2HI\ (2\ \text{torr})$$

(e) AND NOW THE FREE ENERGY

Free energy can be simply calculated from the data in Figures 5-2 and 5-3, since $G = H - TS$ at each value of x. Figure 5-4 shows such a plot, in company with the enthalpy plot of Figure 5-2 and the entropy plot of Figure 5-3 re-expressed to include the temperature multiplier. Now both terms, H and TS, are given in kilocalories, so their relative importance can be assessed.

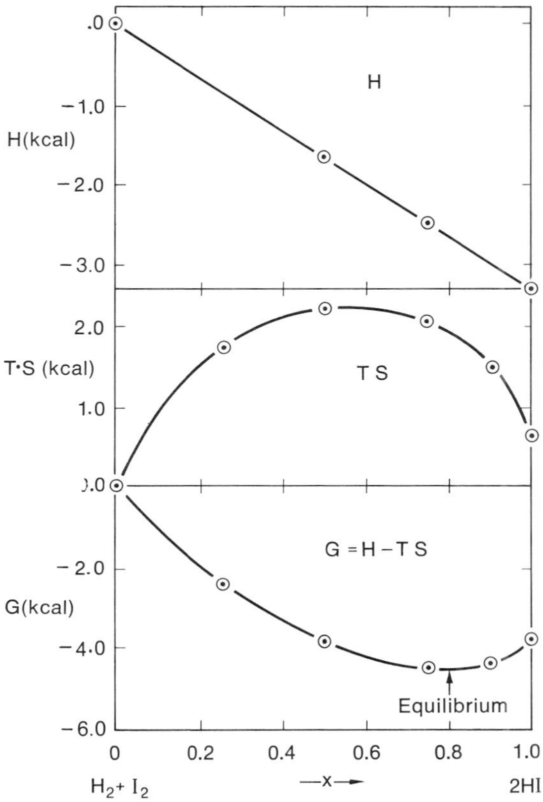

Figure 5-4 Enthalpy (H), entropy (TS) and free energy (G) for the hydrogen–iodine reaction.

The most characteristic feature of *G* is that it has a minimum at $x = 0.81$. This means that an I_2, H_2, HI mixture, one with *x* below 0.81, will react to produce HI, since free energy is moving downhill in that direction. If the mixture initially corresponds to *x* greater than 0.81, the reaction will proceed in the opposite direction, consuming HI and forming the elementary substances H_2 and I_2. Again, that is the downhill direction.

At the minimum of the free energy curve, there will be no net tendency to react in either direction. *At the free energy minimum, equilibrium exists.*

5-2 Free energy—the reaction direction signpost

Knowing that reactions proceed downhill on the free energy landscape, it is clear that free energy will help us make predictions about chemistry. We can answer such a question as "Is ammonia a potential fuel?" If we can learn the free energy change associated with the combustion reaction, its sign will tell us whether the path is downhill. If ΔG is large and negative, the reaction tendency favors products and ammonia may be a useful fuel. The reaction must also be rapid, of course, at an attainable temperature, so thermodynamics doesn't tell us everything we need to know. It also cannot tell us the economics of the process. Nevertheless, thermodynamics is a logical place to start, for *thermodynamics defines the possible.*

(a) STANDARD FREE ENERGIES OF FORMATION

The results of equilibrium and calorimetric studies are extensively tabulated. One such tabulation, used internationally, is the "Selected Values of Chemical Thermodynamic Properties" published by the U.S. National Bureau of Standards.* In this valuable collection are found quantities called the "standard free energy of formation" and "standard enthalpy of formation." Table 5-1 lists some entries from these tables that are relevant to our question about the fuel potentialities of ammonia.

Table 5-1 Standard Enthalpies and Free Energies of Formation ($T = 298.16°K$: 25°C)

	ΔH_f° **(kcal/mole)**	ΔG_f° **(kcal/mole)**
N_2(g, 1 atm)	(0.000)	(0.000)
O_2(g, 1 atm)	(0.000)	(0.000)
H_2(g, 1 atm)	(0.000)	(0.000)
H_2O(g, 1 atm)	−57.7979	−54.6357
NH_3(g, 1 atm)	−11.04	−3.976
NO(g, 1 atm)	21.600	20.719
NO_2(g, 1 atm)	8.091	12.390
N_2O_4(g, 1 atm)	2.309	23.491
N_2O(g, 1 atm)	19.49	24.76

Obviously some advance knowledge was needed to select just nine compounds from the thousands of entries in the 822 page document. Merely to open the reference book meaningfully, it was necessary to write down some possible reactions. Drawing upon previous knowledge of combustion processes, probably learned from a textbook, but possibly by experiment, we decided to consider the following possibilities:

$$NH_3(g) + \tfrac{7}{4}O_2(g) \rightarrow NO_2(g) + \tfrac{3}{2}H_2O(g) \qquad (5\text{-}24)$$

$$NH_3(g) + \tfrac{5}{4}O_2(g) \rightarrow NO(g) + \tfrac{3}{2}H_2O(g) \qquad (5\text{-}25)$$

*"Selected Values of Chemical Thermodynamic Properties, Part I: Tables," National Bureau of Standards Circular 500—Part I, reprinted July 20, 1961. This reference has recently been superceded, in part, by National Bureau of Standards Technical Note 270-3, January 1968.

$$NH_3(g) + O_2(g) \rightarrow \tfrac{1}{2}N_2O(g) + \tfrac{3}{2}H_2O(g) \qquad (5\text{-}26)$$

Each of these three reactions is a candidate for our consideration. In each, ammonia reacts with oxygen. Which one might be important? Perhaps the data in Table 5-1 will tell.

We find in the tables that the thermodynamic properties of each molecule are expressed relative to the elements of which it is composed. Just as in our consideration of the hydrogen–iodine reaction, it is a matter of convenience to choose a reference point for ΔH and ΔG. Convenience alone dictated the choice that chemists have agreed upon: the elements as they exist at 25°C and at pressures of one atmosphere. It is a matter of necessity, however, to have some reference point. *Only changes in enthalpy and free energy are known* (or needed, for a chemist's purposes). Since the elements are selected for reference, they are *assigned* enthalpies and free energies of zero. (Note that our use of 700°C and 1 torr pressure was convenient for the problem in Section 5-1—but would not generally be so.)

The use of these tabulated values depends upon the fact that enthalpy and free energy are both functions of state. This means that the change of enthalpy (or free energy) in a change of state is independent of the path (see Section 2-2 (g) if you wish to brush up on this point). Because of this property, we can consider a chemical change in a step-wise way if we wish. Thus, the change (5-24) can be considered in two steps as shown:

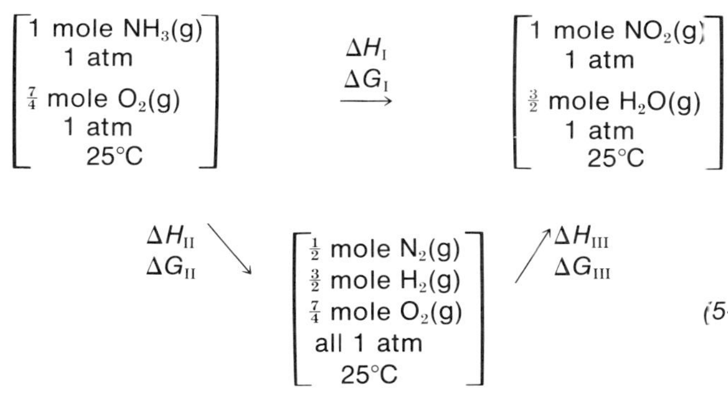

(5-27)

If ΔH_I depends only on the initial and final state, it can be found by summing ΔH_{II} and ΔH_{III}, the enthalpy changes connected with forming the elements from the reactants followed by combining the elements into the products. Thus we can write

$$\Delta H_I = \Delta H_{II} + \Delta H_{III} \qquad (5\text{-}28)$$

We note that ΔH_{III} is merely the sum of the enthalpies of formation of one mole of NO_2 and $\frac{3}{2}$ moles of H_2O from the elements.

The quantity ΔH_{II} is the *negative* of the heat of formation of one mole of NH_3. Hence we obtain

$$\Delta H_I = [-\Delta H_f^0(NH_3)] + [\Delta H_f^0(NO_2) + \tfrac{3}{2}\Delta H_f^0(H_2O)] \tag{5-29}$$

From Table 5-1,

$$\begin{aligned}\Delta H_I &= [-(-11.04)] + [8.091 + \tfrac{3}{2}(-57.7979)] \\ &= -67.56 \text{ kcal}\end{aligned}$$

We conclude that reaction (5-24) is exothermic by 67.56 kcal per mole of NH_3 consumed. Surely this suggests that reaction (5-24) is a possible energy (heat) source. We determined this merely by consulting a reference table of enthalpies of formation and by using the "function of state" property. Thus *any* reaction enthalpy change can be calculated in a few minutes if the tabulated enthalpies of formation are available.

Thermodynamics can also tell us the reaction tendency, the information contained in ΔG. If ammonia is to be used as a fuel, it must not only produce heat, but also, the reaction must occur spontaneously. We learned in Chapter Four that reactions which occur at constant pressure and constant temperature will occur spontaneously if ΔG is negative.

A tabulation of free energies of formation makes the ΔG for any reaction easily accessible. This is because *free energy is also a function of state*. It is such because each of its two parts, ΔH and $T\Delta S$, are themselves functions of state. (Recall that H was shown to be a function of state by the law of energy conservation. Entropy S was defined so as to be a function of state, determined solely by the intrinsic probabilities of initial and final states). The sum (or difference) of two functions of state (as in $G = H - TS$) must also be a function of state.

Referring to the diagram (5-27) and to Table 5-1, we obtain

$$\begin{aligned}\Delta G_I &= \Delta G_{II} + \Delta G_{III} \\ &= [-\Delta G_f^0(NH_3)] + [\Delta G_f^0(NO_2) + \tfrac{3}{2}\Delta G_f^0(H_2O)] \\ &= [-(-3.976)] + [+12.390 + \tfrac{3}{2}(-54.6357)] \\ &= -65.59 \text{ kcal/mole of } NH_3\end{aligned} \tag{5-30}$$

We see that ΔG^0 for reaction (5-24) is negative. Therefore, the reaction between ammonia and oxygen to form nitrogen dioxide and water vapor can proceed spontaneously. If it does, it will release 67.56 kilocalories of heat per mole of ammonia burned. We have said "if it does" because thermodynamics does not tell us the rate at which spontaneous changes occur. In fact, we all know that ammonia does not burst into flame when it is used as a household cleaner, although it is exposed to oxygen in the air. The reaction is too slow. You may not know, however, that ammonia can be passed into an ordinary bunsen burner instead of methane and, once lit, the ammonia burns quite like a normal burner fuel.

In this manner, we can calculate ΔH° and ΔG° for each of the three possible combustion reactions (5-24), (5-25) anc (5-26). The results are compiled in Table 5-2. All three reactions are exothermic, and each one has a tendency to proceed spontaneously, since all three ΔG°'s are negative! Yet we note again that ammonia does not burst into flame on exposure to air at room temperature. There must be impeding energy barriers that cause all three reactions to be extremely slow at room temperature (Fig. 5-5). It is found by experiment that when ammonia does burn, at a high temperature at which reaction rates are high, NO_2 and NO (but not N_2O) are formed in a mixture, the relative amounts

Figure 5-5 Standard free energy signposts: Downhill is the way to go.

Table 5-2 Enthalpy and Free Energy Changes for Possible Ammonia Combustion Reactions (all constituents at 298.16°K and 1 atm pressure)

	ΔH° (kcal)	ΔG° (kcal)
$NH_3(g) + \frac{7}{4}O_2(g) \rightarrow NO_2(g) + \frac{3}{2}H_2O(g)$	−67.56	−65.59
$NH_3(g) + \frac{5}{4}O_2(g) \rightarrow NO(g) + \frac{3}{2}H_2O(g)$	−54.06	−57.26
$NH_3(g) + O_2(g) \rightarrow \frac{1}{2}N_2O(g) + \frac{3}{2}H_2O(g)$	−65.90	−65.60

of each being dependent upon the amount of oxygen available. Thermodynamics tells us (in Table 5-2) that any one of the three reactions (5-24), (5-25), or (5-26) *could* occur and, if it did, heat would be released. However, it cannot tell us the speed of reaction for any of the three.

5-3 Free energy and equilibrium

At first glance, Sections 5-1 and 5-2 seem to be contradictory. In Section 5-1 we found not only that the reactants H_2 and I_2 can react spontaneously, running downhill on the G terrain, but also that pure HI can react spontaneously, again running downhill. The minimum free energy exists when *both* reactants and products are present. On the other hand, in Section 5-2 we tried to answer our question about ammonia merely by asking whether ΔG is positive or negative. We shall see that these two discussions are quite compatible, as long as we know what they mean.

(a) REACTION UNDER STANDARD CONDITIONS

Reaction (5-27) specifies plainly that the change considered is that of one mole of ammonia and 7/4 moles of oxygen reacting at fixed pressures, each at one atmosphere. Furthermore, the products are formed at fixed pressures, each at one atmosphere. To carry out such a reaction in the laboratory, all the reactants and products would have to be mixed in a reaction vessel, each at one atmosphere pressure. For this "standard" reaction mixture, the free energy change is called the "standard free energy" and it is designated ΔG°. The algebraic sign of the "standard" free energy tells us the "direction" chemical change will take from these starting conditions—whether reactants, each at one atmosphere, will be consumed at 298°K to form products at one atmosphere, or vice versa.

Figure 5-4 seems to be more informative. It displays the fact that there is a particular mixture for which free energy is a minimum. If we begin with an equimolar mixture of H_2 and I_2 (each at 1 torr pressure), we find that reaction will proceed until the mole fraction of HI is 0.81. If we begin with pure HI (at 2 torr

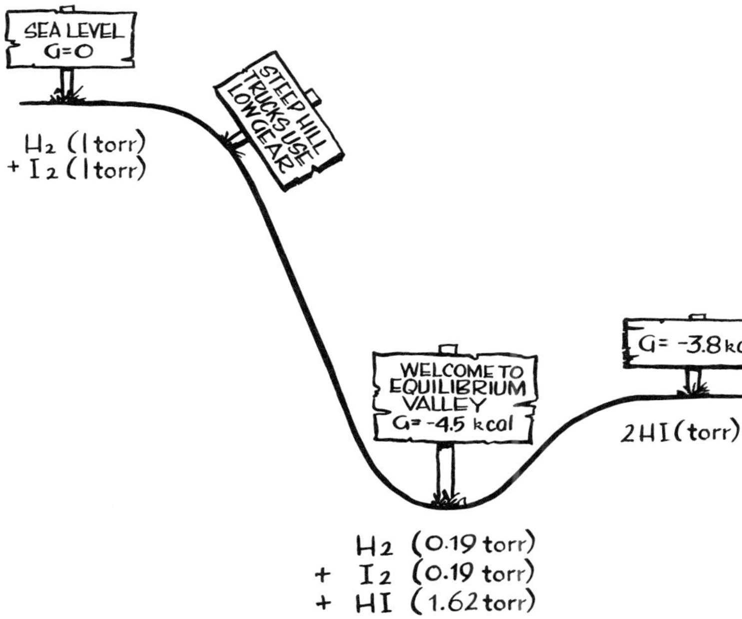

Figure 5-6 Down the free energy hill to equilibrium valley.

pressure), it will decompose to form H_2 and I_2, but again only until the mole fraction of HI is 0.81 (Fig. 5-6). At this composition, equilibrium exists—there remains no net tendency for change.

Plainly, we need to know as much as possible about equilibrium. ΔG° merely "points" towards equilibrium, the free energy minimum. We would like to know how far away it is. Surprisingly, it turns out that ΔG° tells us that, as well. Here is how the story goes.

(b) REACTION UNDER NON-STANDARD PRESSURE CONDITIONS

Given ΔG°, the free energy change under standard concitions (1 atm, 25°C), it would be helpful to be able to calculate ΔG for some other set of conditions that might interest us. For the moment, let's consider changing only the pressures. For example, consider the oxidation of nitric oxide NO to nitrogen dioxide NO_2.

$$\underset{\substack{1\text{ atm}\\25^{\circ}\text{C}}}{\text{NO(g)}} + \underset{\substack{1\text{ atm}\\25^{\circ}\text{C}}}{\tfrac{1}{2}\text{O}_2\text{(g)}} \xrightarrow[\Delta G^{\circ}]{\Delta H^{\circ}} \underset{\substack{1\text{ atm}\\25^{\circ}\text{C}}}{\text{NO}_2\text{(g)}} \tag{5-31}$$

Table 5-1 contains the necessary information for us to calculate the free energy and enthalpy changes under these standard conditions: $\Delta H^\circ = -13.51$ kcal and $\Delta G^\circ = -8.33$ kcal. The reaction is exothermic and has a tendency to react as shown.

Now suppose our interest is in the reaction as written except for one change—we would like to consider the conditions under which NO_2 is formed at a pressure of *one-half* atmosphere.

$$\begin{array}{ccccc} NO(g) & + \tfrac{1}{2}O_2(g) & \xrightarrow[\Delta G]{\Delta H} & NO_2(g) \\ 1 \text{ atm} & 1 \text{ atm} & & \tfrac{1}{2} \text{ atm} \\ 25^\circ C & 25^\circ C & & 25^\circ C \end{array} \qquad (5\text{-}32)$$

In determining ΔH and ΔG for such non-standard conditions, the traditional strategy is to try to devise a stepwise way of accomplishing the change. If a path can be found for which ΔH and ΔG for each step are easily learned, the overall ΔH and ΔG can be evaluated, since they are independent of path. In our case, we would like a stepwise process that accomplishes (5-32), but includes as one of its steps the change under standard conditions, since we know ΔH° and ΔG°. Here is such a two-step process.

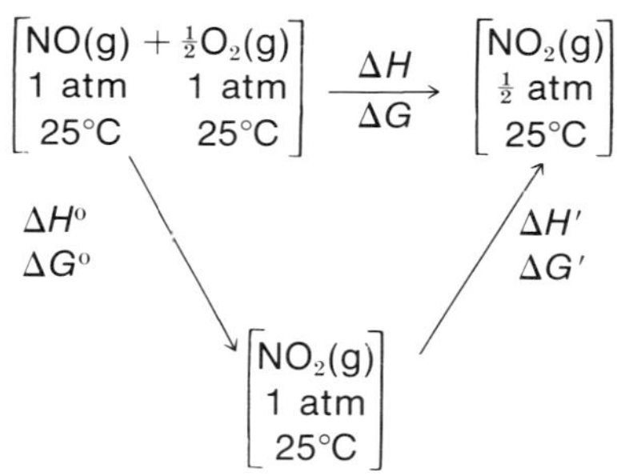

(5-33)

Plainly,

$$\Delta H = \Delta H^\circ + \Delta H' \qquad (5\text{-}34)$$

and

$$\Delta G = \Delta G^\circ + \Delta G' \qquad (5\text{-}35)$$

We have already looked up ΔH° and ΔG° in the tables, so all we need to do is calculate $\Delta H'$ and $\Delta G'$.

The first quantity $\Delta H'$ is the easier of the two. Remember that

$$H = E + PV$$

so

$$\Delta H' = \Delta E' + \Delta(PV)'$$

But for a perfect gas, E is a function of temperature only. This means that the energy content of a perfect gas is constant at a fixed temperature, so $\Delta E' = 0$. Also for a perfect gas $PV = nRT$ and, if T is constant, the PV product is constant. So $\Delta(PV)' = 0$, and hence,

$$\Delta H' = 0$$

$$\boxed{\Delta H = \Delta H^{\circ}} \tag{5-36}$$

The second quantity $\Delta G'$ is more interesting. Since $G = H - TS$,

$$\Delta G' = \Delta H' - \Delta(TS)' \tag{5-37}$$

Substituting (5-36) into (5-37) and remembering that T is constant, (5-37) becomes

$$\Delta G' = 0 - \Delta(TS)' = -T\Delta S' \tag{5-38}$$

The quantity $\Delta S'$ is just the entropy change accompanying a constant temperature expansion. We already dealt with that situation in Section 5-1 (d). There the entropy was shown to be $-nR \log_e(P_2/P_1)$. Now we have

$$\Delta G' = -T\Delta S' = -T\left(-nR \log_e \frac{P_2}{P_1}\right)$$
$$= nRT \log_e \frac{P_2}{P_1} \tag{5-39}$$

and

$$\Delta G = \Delta G^{\circ} + \Delta G'$$

so

$$\Delta G = \Delta G^{\circ} + nRT \log_e \frac{P_2}{P_1} \tag{5-40}$$

Expression (5-40) can be simplified even further because P_1 is the standard pressure, one atmosphere, and $\log_e 1 = 0$, so

$$\Delta G = \Delta G^{\circ} + nRT \log_e \frac{P_2}{1} = \Delta G^{\circ} + nRT(\log_e P_2 - \log_e 1)$$

$$\boxed{\Delta G = \Delta G^{\circ} + nRT \log_e P_2} \tag{5-41}$$

In (5-40), n refers to the number of moles of NO_2, the gas that expanded to P_2.

(c) REACTION UNDER EQUILIBRIUM CONDITIONS

It is clear that we can compute ΔG for *any* set of pressure conditions exactly as we did (5-41). For example, consider reaction (5-31) occurring at generalized pressures p_{NO}, p_{O_2}, and p_{NO_2}. Our multistep process would be the following:

$$\begin{array}{ccccc} \begin{bmatrix} NO(g) \\ p_{NO} \\ 25°C \end{bmatrix} & + & \begin{bmatrix} \frac{1}{2}O_2(g) \\ p_{O_2} \\ 25°C \end{bmatrix} & \xrightarrow[\Delta G]{\Delta H} & \begin{bmatrix} NO_2(g) \\ p_{NO_2} \\ 25°C \end{bmatrix} \\ \Delta H_1 \downarrow \Delta G_1 & & \downarrow \Delta H_2, \Delta G_2 & & \uparrow \Delta H_3, \Delta G_3 \\ \begin{bmatrix} NO(g) \\ 1\text{ atm} \\ 25°C \end{bmatrix} & + & \begin{bmatrix} \frac{1}{2}O_2(g) \\ 1\text{ atm} \\ 25°C \end{bmatrix} & \xrightarrow[\Delta G^\circ]{\Delta H^\circ} & \begin{bmatrix} NO_2(g) \\ 1\text{ atm} \\ 25°C \end{bmatrix} \end{array} \qquad (5\text{-}42)$$

Thus

$$\Delta H = \Delta H_1 + \Delta H_2 + \Delta H^\circ + \Delta H_3 \qquad (5\text{-}43)$$

and

$$\Delta G = \Delta G_1 + \Delta G_2 + \Delta G^\circ + \Delta G_3 \qquad (5\text{-}44)$$

Just as in (5-36), at constant temperature, $\Delta H_1 = \Delta H_2 = \Delta H_3 = 0$, so

$$\boxed{\Delta H = \Delta H^\circ} \qquad (5\text{-}45)$$

The quantities ΔG_1, ΔG_2, and ΔG_3 are all evaluated as easily as (5-39).

$$\Delta G_1 = n_{NO}\, RT \log_e \frac{1\text{ atm}}{p_{NO}}$$

$$\Delta G_2 = n_{O_2}\, RT \log_e \frac{1\text{ atm}}{p_{O_2}}$$

and

$$\Delta G_3 = n_{NO_2}\, RT \log_e \frac{p_{NO_2}}{1\text{ atm}}$$

hence,

$$\Delta G = \Delta G^\circ + n_{NO}RT(-\log_e p_{NO}) + n_{O_2}RT(-\log_e p_{O_2}) + n_{NO_2}RT(\log_e p_{NO_2})$$

or, since $n \log p = \log p^n$,

$$\Delta G = \Delta G^\circ - RT \log_e(p_{NO})^{n_{NO}} - RT \log_e(p_{O_2})^{n_{O_2}} + RT \log_e(p_{NO_2})^{n_{NO_2}}$$

that is,

$$\Delta G = \Delta G^{\circ} + RT \log_e \frac{(p_{NO_2})^{n_{NO_2}}}{(p_{NO})^{n_{NO}}(p_{O_2})^{n_{O_2}}}$$

Inserting $n_{NO_2} = 1$, $n_{NO} = 1$, and $n_{O_2} = \frac{1}{2}$ from the balanced ecuation (5-32),

$$\boxed{\Delta G = \Delta G^{\circ} + RT \log_e \frac{p_{NO_2}}{p_{NO}\, p_{O_2}^{1/2}}} \quad (5\text{-}46)$$

Equation (5-46) can be applied to calculate ΔG for any set of NO_2, NO, and O_2 pressures. For example, it can be applied at any composition on a free energy plot, such as the one shown in Figure 5-4. In fact it can even be applied at the minimum in the free energy curve, the point at which equilibrium exists. Under these equilibrium conditions, a minute change either towards reactants or towards products causes no change in free energy. At equilibrium, $\Delta G = 0$. This special relationship really makes equation (5-46) take on its greatest meaning. When the gas pressures reach equilibrium values,

$$\Delta G = 0 = \Delta G^{\circ} + RT \log_e \frac{(p_{NO_2})_{eq}}{(p_{NO})_{eq}(p_{O_2})_{eq}^{1/2}}$$

so

$$\Delta G^{\circ} = -RT \log_e \frac{(p_{NO_2})_{eq}}{(p_{NO})_{eq}\,(p_{O_2})_{eq}^{1/2}} \quad (5\text{-}47)$$

Expression (5-47) can be rearranged as follows:

$$\log_e \frac{(p_{NO_2})_{eq}}{(p_{NO})_{eq}\,(p_{O_2})_{eq}^{1/2}} = -\frac{\Delta G^{\circ}}{RT} \quad (5\text{-}48)$$

or

$$\frac{(p_{NO_2})_{eq}}{(p_{NO})_{eq}\,(p_{O_2})_{eq}^{1/2}} = e^{-\Delta G^{\circ}/RT} \quad (5\text{-}49)$$

On the left-hand side in (5-49), we see a relation among concentrations (pressures) of reactants and products—the familiar equilibrium expression! On the right, we find an exponential, the exponent of which is fixed by ΔG° and T. Since ΔG° is just a number characteristic of the reaction, $\Delta G^{\circ}/RT$ is a constant at any given temperature. Hence the right-hand side of (5-49) must be a constant, and this constant must be the equilibrium constant! Hence (5-49) can be rewritten as

$$\frac{(p_{NO_2})_{eq}}{(p_{NO})_{eq}\,(p_{O_2})_{eq}^{1/2}} = K = e^{-\Delta G^{\circ}/RT} \quad (5\text{-}50)$$

and (5-48) becomes

$$\boxed{\Delta G^{\circ} = -RT \log_e K = -2.303RT \log_{10} K} \qquad (5\text{-}51)$$

Expression (5-50) is the familiar equilibrium law, the one we obtained in Chapter One through reaction-rate arguments. Now we have seen its origin in thermodynamics. Expression (5-51) tells us how to calculate K from standard free energies. Obviously these are extremely valuable relationships.

(d) THE ΔG REACTION SIGNPOST AT WORK

We might exercise relations (5-46) and (5-50) by applying them to the H_2–I_2 reaction explored in Section 5-1. First, ΔG° is needed and, once again, we can obtain this from the standard free energies of formation of the reactants and products, as given in the National Bureau of Standards tables.

$$H_2(g) + I_2(g) = 2HI(g) \qquad (5\text{-}52)$$

$$\begin{aligned} \Delta G^{\circ} &= 2\Delta G^{\circ}_{f}(HI) - \Delta G^{\circ}_{f}(H_2) - \Delta G^{\circ}_{f}(I_2) \\ &= 2(0.31) - (0.00) - (4.63)^{*} \\ &= -4.01 \text{ kcal} \end{aligned} \qquad (5\text{-}53)$$

Now we can apply (5-51) to calculate the equilibrium constant at 298.16°K for reaction (5-52):

$$\begin{aligned} K &= e^{-\Delta G^{0}/RT} = e^{-(-4010)/(1.99\cdot 298)} = e^{+6.76} \\ &= 10^{+6.76/2.303} = 10^{+2.93} \\ &= 8.6 \cdot 10^{+2} \end{aligned} \qquad (5\text{-}54)$$

With this equilibrium constant, we can calculate the equilibrium composition that would result if H_2 gas and I_2 gas, each at 1 torr pressure, were to react at room temperature (perhaps in the presence of a catalyst). In terms of our quantity x, the equilibrium pressures would be

$$(p_{H_2})_{eq} = 1 - x$$
$$(p_{I_2})_{eq} = 1 - x$$
$$(p_{HI})_{eq} = 2x$$

These can now be substituted in the equilibrium expression

$$K = \frac{(p_{HI})^2_{eq}}{(p_{H_2})_{eq}(p_{I_2})_{eq}}$$

$$= \frac{(2x)^2}{(1-x)^2} \qquad (5\text{-}55)$$

*Note that in the tables, the standard free energy of formation of gaseous I_2 is not zero because the standard state is taken to be *solid* iodine at 298.16°K. The $\Delta G^{\circ}_{f}(I_2)$ given, refers to the hypothetical iodine gas at one atmosphere pressure formed from solid iodine, the reference state.

The solution to this expression is $x = 0.94$. At room temperature, 94 percent of the H_2 will react to reach the equilibrium situation. Remember from Figure 5-4 that at the higher temperature 700°K, the equilibrium mixture involved an 81 percent conversion. This is the appropriate direction of change, as suggested by LeChatelier's Principle, which indicates that since the reaction releases heat, a rise in temperature will shift equilibrium in the endothermic direction, towards reactants. Thermodynamics tells us why and exactly how much.

Now we might apply relation (5-46) at other than equilibrium conditions; for example, on either side of equilibrium. Let's begin with $x = 0.50$, so that $p_{H_2} = p_{I_2} = 0.50$ torr and $p_{HI} = 1.00$ torr. Now

$$\begin{aligned}\Delta G_{x=0.50} &= \Delta G^\circ + RT \log_e \frac{(1.00)^2}{(0.5)(0.5)} \\ &= -4010 + (2.303)(1.99)(298.16) \log_{10} 4 \\ &= -4010 + 820 \\ &= -3190 \text{ cal}\end{aligned}$$

We see that at $x = 0.50$, ΔG is still negative, so equilibrium lies ahead, in the direction of more reaction. If we now consider $x = 0.95$, so that $p_{H_2} = p_{I_2} = 0.05$ torr and $p_{HI} = 1.90$ torr, the calculation changes to

$$\begin{aligned}\Delta G_{x=0.95} &= \Delta G^\circ + RT \log_e \frac{(1.90)^2}{(0.05)(0.05)} \\ &= -4010 + (2.303)(1.99)(298.16) \log_{10} 1444 \\ &= -4010 + 4310 \\ &= +300 \text{ cal}\end{aligned}$$

Now ΔG is positive, so our direction signpost says that equilibrium lies behind us. At $x = 0.95$, the downhill direction is back towards reactants.

We see that the free energy change ΔG° tells us the direction of equilibrium beginning with the very special (and usually uninteresting) mixture of all reactants and products at the same pressure of one atmosphere. But ΔG, calculated via expression (5-46) tells us the direction of equilibrium from *any* starting mixture. If ΔG turns out to be negative, reaction can proceed toward products. If ΔG turns out to be positive, the mixture is already beyond equilibrium, and it can only proceed in the reverse direction, consuming the products and producing the reactants expected from the reaction as written.

5-4 Temperature dependence of ΔH, ΔS, and ΔG

The temperature dependencies of thermodynamic properties are obviously important. Most important is the fact that equilibrium conditions are observed to change rapidly with temperature, implying that temperature provides a powerful measure of control over chemical reactions. We shall explore these dependencies.

(a) ENTHALPY AND ENTROPY—TEMPERATURE INDEPENDENT (ALMOST)

Experimentally, we find that for most reactions, neither ΔH nor ΔS changes dramatically with temperature. Table 5-3 shows this for three typical gas-phase reactions and a couple of typical reactions involving solids. The first three columns show the enthalpy change at the standard temperature 298°K, the change at twice that temperature, 596°K, and the percentage difference in ΔH. The last three columns do the same for entropy change.

In the typical examples given, the value of ΔH changes by, at most, a few percent as the absolute temperature doubles. *To a very good and useful approximation, we can say that ΔH is temperature independent.*

The value of ΔS changes rather more—up to 10 percent for the ethylene–hydrogen reaction (C_2H_4–H_2). Nevertheless, the most striking property of ΔS is that it changes characteristically only a few percent per hundred degree change in temperature, making it unimportant for most considerations.

It is not too difficult to perceive why these two thermodynamic quantities should be relatively insensitive. The enthalpy change and the entropy change express the energy and randomness properties of the reaction products *relative* to the reactants. When the temperature is raised, the enthalpy of the products increases but so also does the enthalpy of the reactants. Only if they go up by different amounts does a change in ΔH occur. The same is true for randomness. While the intrinsic randomness of the reactants increases as the temperature rises (there is more motional

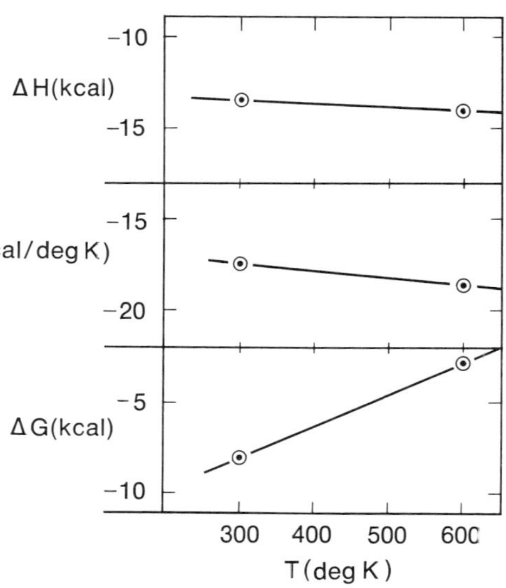

Figure 5-7 Temperature dependence of thermodynamic quantities
$NO(g) + \frac{1}{2}O_2(g) = NO_2(g)$

Table 5-3 Temperature Dependence of ΔH and ΔS

Reactions involving gases	ΔH(298°K) (kcal)	ΔH(596°K) (kcal)	Percentage change	ΔS(298°K) (cal/degK)	ΔS(596°K) (cal/degK)	Percentage change
$NO(g) + \frac{1}{2}O_2(g) = NO_2(g)$	−13.5	−14.0	3.5	−17.4	−18.5	6.3
$CO(g) + \frac{1}{2}O_2(g) = CO_2(g)$	−67.6	−68.1	0.7	−20.7	−21.8	5.3
$C_2H_4(g) + H_2(g) = C_2H_6(g)$	−32.7	−34.0	3.8	−28.8	−31.6	9.7
Reactions involving solids						
$CaCO_3(s) = CaO(s) + CO_2(g)$	+42.6	+42.1	1.1	+38.4	+37.4	2.7
$C(s) + \frac{1}{2}O_2(g) = CO(g)$	−26.4	−26.0	1.6	+21.4	+22.4	4.5

randomness), it also increases for the products. The net change in ΔS is the difference between these two values, and, as for ΔH, there is a tendency for these changes to be of similar magnitude and therefore to cancel.

In conclusion, *it is a useful approximation to assume that ΔH and ΔS are temperature independent.* For temperature changes of the order of 100°, ΔH remains unchanged to less than one percent, and ΔS usually changes no more than two or three percent.

(b) FREE ENERGY CHANGE WITH TEMPERATURE—ANOTHER STORY

Table 5-4 shows ΔG for the same reactions listed in Table 5-3. Here, the *smallest* change is 10 percent, and ΔG for nitric oxide changes by a factor of 2.8. For some reactions, a temperature change can actually change the sign of ΔG from positive to negative. The free energy is by no means a temperature-independent quantity.

Table 5-4 Temperature Dependence of ΔG

	ΔG(298°K) (kcal)	ΔG(596°K) (kcal)	Percentage change
Reactions involving gases			
$NO(g) + \frac{1}{2}O_2(g) = NO_2(g)$	−8.33	−2.97	+64
$CO(g) + \frac{1}{2}O_2(g) = CO_2(g)$	−61.4	−55.1	+10.3
$C_2H_4(g) + H_2(g) = C_2H_6(g)$	−24.1	−15.2	+37
Reactions involving solids			
$CaCO_3(s) = CaO(s) + CO_2(g)$	+31.2	+19.8	−36.6
$C(graphite) + \frac{1}{2}O_2(g) = CO(g)$	−32.8	−39.4	−20.0

Again it is readily seen why ΔG changes rapidly with temperature, although ΔH and ΔS do not. Remembering the definition of ΔG for a reaction occurring at a fixed temperature T_1, the free energy change will be

$$\Delta G_1 = \Delta H - T_1\Delta S \qquad (5\text{-}56)$$

At a different temperature T_2, ΔG will become

$$\Delta G_2 = \Delta H - T_2\Delta S \qquad (5\text{-}57)$$

Expressions (5-56) and (5-57) show that the total randomness contribution to ΔG changes in proportion to temperature because of the T factor in $T\Delta S$. In fact, to a good approximation, a plot of ΔG against temperature gives a straight line. Figure 5-8 shows typical data for the reaction

$$M^+(FHF)^-(s) \rightleftarrows M^+F^-(s) + HF(g) \qquad (5\text{-}58)$$

Obviously a straight line fits the experimental points quite well.

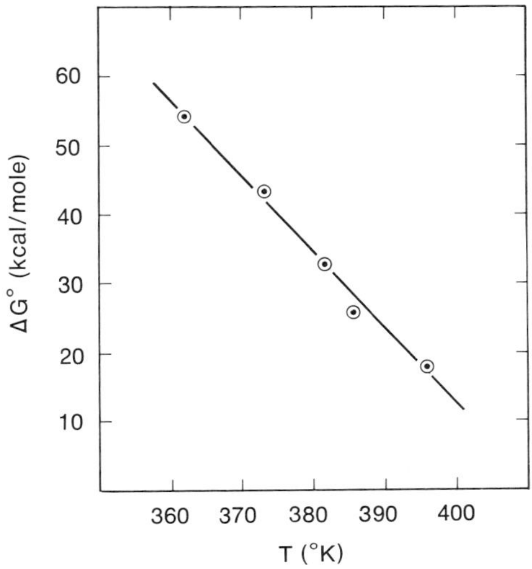

Figure 5-8 Variation of ΔG° *with temperature for the dissociation reaction*

$M^+(FHF)^-(s) \rightleftharpoons M^+F^-(s) + HF(g)$

(The abbreviation M^+ *is used for the tetramethyl ammonium ion* $(CH_3)_4N^+$*.)*

(c) ΔG° AGAINST T: AN EASY MEASUREMENT OF ΔH°

The standard free energy is related simply to the equilibrium constant (as shown in Section 5-4):

$$\Delta G^\circ = -RT \log_e K = -2.303\ RT \log_{10} K \qquad (5\text{-}51)$$

Consequently, expressions (5-56) and (5-57) can be rewritten in terms of the equilibrium constants K_1 and K_2 (at temperatures T_1 and T_2), provided the ΔH and ΔS values refer to standard conditions (1 atmosphere pressure for every constituent).

section 5-4 temperature dependence of ΔH, ΔS, and ΔG

$$-RT_1 \log_e K_1 = \Delta H^\circ - T_1 \Delta S^\circ \qquad (5\text{-}59)$$

and

$$-RT_2 \log_e K_2 = \Delta H^\circ - T_2 \Delta S^\circ \qquad (5\text{-}60)$$

A more convenient form is obtained by dividing by $-RT$,

$$\log_e K_1 = -\frac{\Delta H^\circ}{RT_1} + \frac{\Delta S^\circ}{R} \qquad (5\text{-}61)$$

and

$$\log_e K_2 = -\frac{\Delta H^\circ}{RT_2} + \frac{\Delta S^\circ}{R} \qquad (5\text{-}62)$$

These two expressions can be combined to relate the temperature change in K to ΔH. Subtracting (5-62) from (5-61), the ΔS° term cancels out completely:

$$\log_e K_1 - \log_e K_2 = -\frac{\Delta H^\circ}{RT_1} + \frac{\Delta H^\circ}{RT_2}$$

or, more tidily,

$$\boxed{\log_e \frac{K_1}{K_2} = -\frac{\Delta H^\circ}{R}\left(\frac{1}{T_1} - \frac{1}{T_2}\right)} \qquad (5\text{-}63)$$

The relationship (5-63) is very important to chemists because it generally provides the easiest possible measurement of ΔH°. The two equilibrium constants can be measured merely through constituent analysis at equilibrium. While this is not child's play and often requires quite sophisticated experiments, it is generally much more reliable than a direct calorimetric measurement of ΔH. It is, of course, obvious that measurement of K gives ΔG° directly through (5-51), and once ΔH° is known, ΔS° can also be calculated from (5-59). Hence, *a measurement of the temperature dependence of the equilibrium constant suffices to give all the thermodynamic functions,* ΔG°, ΔH°, ΔS°, *over that temperature range.*

5-5 Conclusion

In this chapter we have seen how the Equilibrium Law can be derived from thermodynamic arguments. The free energy change accompanying a reaction tends toward a minimum, and, at this minimum, equilibrium prevails. The equilibrium constant that expresses the equilibrium conditions depends upon ΔG°, the free energy change under standard conditions. Finally, the temperature dependence of the equilibrium constant provides a convenient measurement of ΔH° and ΔS°.

We can now see the great importance of thermodynamic considerations in chemistry. The motivation for spontaneous change is the tendency to maximum randomness. For the familiar "open beaker" experiments conducted at constant pressure and constant temperature, this tendency to maximum randomness in the universe is connected with a tendency to minimum free energy in the system. Hence, the direction and distance to equilibrium are measured by ΔG. Finally, the heat effects that accompany such a change are measured by ΔH. In the next chapter, we will see these powerful principles at work in a wide variety of important applications.

We now have before us the important conceptual ideas of thermodynamics. In this concluding chapter, we will consider some applications of these ideas to particularly important systems. First, electrochemical cells will again be considered. Then the special importance of entropy effects in aqueous solutions will be explored. After that, absolute entropies and the Third Law of Thermodynamics will be briefly considered. Finally, we will deal, in a rather qualitative way, with the application of thermodynamics to biological processes—the opening field called bioenergetics.

six chemical thermodynamics in action

6-1 Electrochemical cells

In Chapter Four, we saw how a spontaneous chemical change could be employed to produce useful work in an electrochemical cell. The amount of work that can be extracted is maximized when the cell is operated reversibly; that is, when the current is minimized. Now that we have the free energy concept at our disposal, it is timely to return to the electrochemical cell and investigate this maximum work.

(a) FREE ENERGY AND MAXIMUM WORK

Electrochemical cells are usually operated in open beakers; hence, at constant pressure and thermostatted at constant temperature. Under such common conditions, the free energy change that accompanies the cell reaction is simply related to the electrical work that can be extracted. This is readily seen by the use of the definitions of $\Delta G, \Delta S,$ and ΔE.

At constant temperature,

$$\Delta G = \Delta H - T\Delta S \qquad (6\text{-}1)$$

and, at constant pressure,

$$\Delta H = \Delta E + P\Delta V \qquad (6\text{-}2)$$

Inserting (6-2) into (6-1),

$$\Delta G = \Delta E + P\Delta V - T\Delta S \qquad (6\text{-}3)$$

Now we recall the relationships that ΔS and ΔE have to reaction heats:

$$\Delta E = q - w \qquad (6\text{-}4)$$

and

$$\Delta S = \frac{q_{\text{rev}}}{T} \tag{6-5}$$

Substituting (6-4) and (6-5) into (6-3),

$$\Delta G = q - w + P\Delta V - q_{\text{rev}}$$

or

$$\Delta G = (q - q_{\text{rev}}) - w + P\Delta V \tag{6-6}$$

Now, in an electrochemical cell, there are two kinds of work. The first is the $P\Delta V$ expansion work associated with pushing back the atmosphere at constant pressure. The second, we will call w_{e}, the electrical work done by the cell. Putting these explicitly into (6-6), we obtain

$$\Delta G = (q - q_{\text{rev}}) - (P\Delta V + w_{\text{e}}) + P\Delta V$$

or

$$\Delta G = (q - q_{\text{rev}}) - w_{\text{e}} \tag{6-7}$$

In Section 4-2, the succession of ways in which the electrochemical cell was discharged showed that there is a maximum amount of electrical work $w_{\text{e,max}}$ that can be obtained. However, this maximum work is obtained only if the cell is discharged infinitely slowly. Under these reversible conditions, q becomes equal to q_{rev}:

$$\Delta G = (q_{\text{rev}} - q_{\text{rev}}) - w_{\text{e,max}}$$

or

$$\boxed{\Delta G = -w_{\text{e,max}}} \tag{6-8}$$

This result is the reason ΔG is called the *free* energy. At constant pressure and temperature, ΔG measures the maximum work other than the pressure–volume expansion work. In an open beaker, this $P\Delta V$ work is wasted; it merely pushes back the atmosphere. The rest of the possible work, measured by ΔG, is "available" or "free" for our use.

(b) MAXIMUM WORK AND CELL VOLTAGE

In Chapter Four, the relation between work performed and cell voltage was developed. Recapitulating, the current I that can flow is limited by the cell voltage $\mathscr{E}$ and the total resistance (internal plus external) R.

$$I = \frac{\mathscr{E}}{R} \text{ coulombs/sec} \tag{6-9}$$

However, the current is a measure of the charge movement per unit time. Therefore, if N electrons flow past a point in time t, the current I is merely

$$I = \frac{Ne}{t}$$

or

$$t = \frac{Ne}{I} = \frac{NeR}{\mathscr{E}} \tag{6-10}$$

Since the power generated by a current I is given by I^2R, and energy is equal to the product of power and time, the energy available to be converted into work is as follows:

$$\begin{aligned} \text{energy available for work} &= (I^2R)(t) \\ &= \left(\frac{\mathscr{E}}{R}\right)^2 \cdot R \cdot t \\ &= \frac{\mathscr{E}^2}{R^2} \cdot R \cdot \frac{NeR}{\mathscr{E}} = Ne\mathscr{E} \end{aligned}$$

Therefore,

$$w_{\text{e,max}} = Ne\mathscr{E} \tag{6-11}$$

This is the form in which we left the electrical work in Chapter Four. It is more convenient, however, to speak in terms of $n = N/N_0$, the number of moles of electrons, and to incorporate both N_0, Avogadro's number, and the electron charge e into a conversion factor $\mathscr{F}$ that adjusts the units to suit our taste.

$$\begin{aligned} w_{\text{e,max}} &= Ne\mathscr{E} \\ &= (nN_0)e\mathscr{E} = n(N_0e)\mathscr{E} \end{aligned}$$

and

$$\boxed{w_{\text{e,max}} = n\mathscr{E}\mathscr{F}} \tag{6-12}$$

where n is the number of moles of electrons exchanged in reaction; $\mathscr{E}$ is the voltage generated by the cell at zero current; $\mathscr{F}$ is equal to 96,500 if the work is expressed in joules, equal to 23,082 if the work is expressed in calories, and equal to unity if the work is expressed in electron volts.

(c) FREE ENERGY AND CELL VOLTAGE

By combining expressions (6-8) and (6-12), we have a powerful conclusion. If

$$\Delta G = -w_{\mathrm{e,max}} \qquad (6\text{-}8)$$

and

$$w_{\mathrm{e,max}} = n\mathscr{E}\mathscr{F} \qquad (6\text{-}12)$$

we have

$$\boxed{\Delta G = -n\mathscr{E}\mathscr{F}} \qquad (6\text{-}13)$$

We seem to have encountered a new way to measure free energy changes! Merely by preparing an electrochemical cell and measuring its voltage under reversible conditions, we can determine ΔG. The expression "under reversible conditions" merely means "zero-current conditions," and this turns out to be quite easily achieved.

Figure 6-1 shows a Zn–Ag cell set in opposition to a fraction of the voltage $\mathscr{E}_0'$ of a standard cell. The fraction is fixed by the ratio of R/R_o, which depends upon the position of the slide contact. For small values of R (Figure 6-1a), the Zn–Ag voltage will exceed the opposing voltage $(R_a/R_o) \cdot \mathscr{E}_0'$, and current will flow in the direction that discharges our cell. For large values of R (Figure 6-1b), the Zn–Ag voltage will be less than the opposing voltage $(R_b/R_o) \cdot \mathscr{E}_0'$, and the electron current will be reversed, flowing so as to recharge our cell. At the slide-contact position at which no current flows at all (Figure 6-1c), the measured voltage $(R_c/R_o) \cdot \mathscr{E}_0$ exactly equals the Zn–Ag voltage. Since the opposition of equal voltages stops current flow through the cell, its voltage is measured under reversible conditions. The ease with which a reaction can be studied under reversible conditions in an electrochemical cell is the reason for the particular charm these devices have for a chemist. *Cell-voltage measurements provide one of the best means of determining free energy changes for reactions that occur in aqueous solutions.* Of course, the absence of electrical conductivity limits the usefulness of such measurements in other solvents and for entirely gas-phase reactions.

(d) STANDARD CELL VOLTAGES (POTENTIALS) AND THE EQUILIBRIUM CONSTANT

If we can determine ΔG through a cell-voltage measurement, we can apply the measurement to the special case of a cell operating

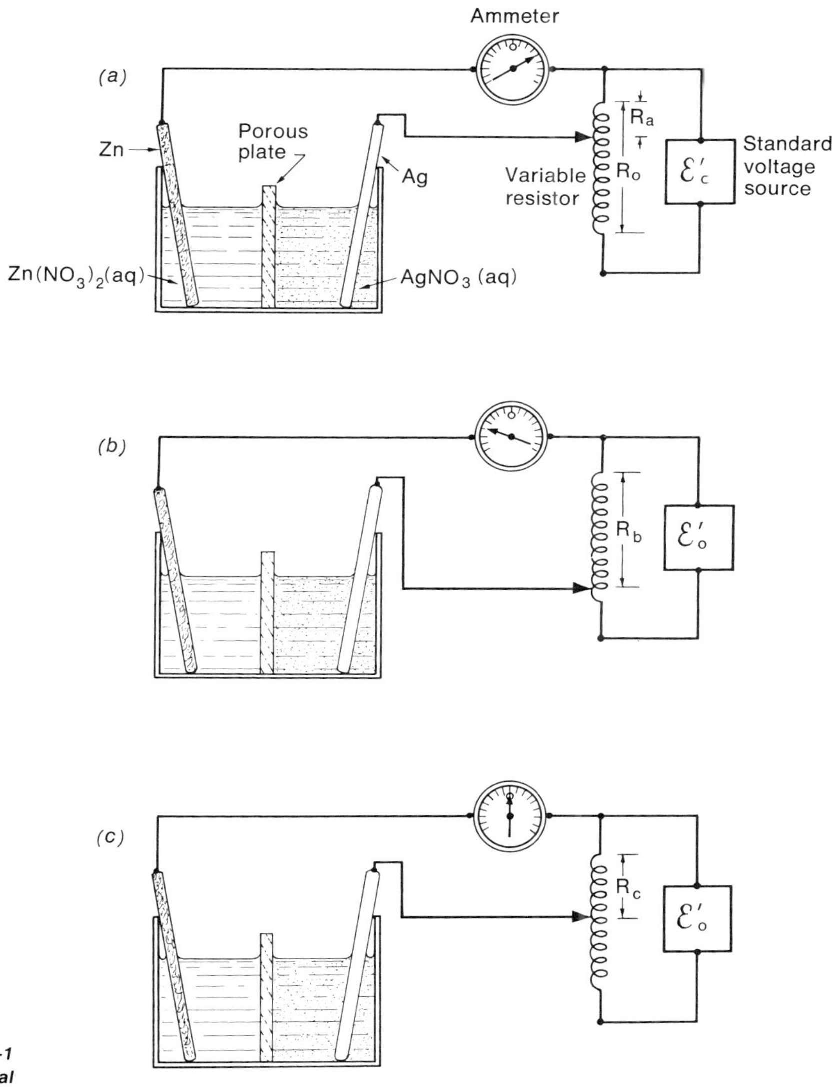

Figure 6-1 Measurement of ℰ of an electrochemical cell.

with all concentrations set at the standard states. This means that all the gases involved should be at one atmosphere pressure, soluble substances should be at one mole per liter concentration,*

*A more rigorous statement would specify "unit activity" instead of "one mole/liter concentration." The "activity" differs from the concentration by a multiplicative factor that corrects for restrictions of mobility due to intermolecular interactions. One manifestation of these mobility restrictions is found in a lack of constancy of an equilibrium constant (when calculated from the Equilibrium Law using measured concentrations). In aqueous solutions, the effects are generally negligible at 10^{-3} M but may be in the range 10–20 percent at 0.1 M.

and solids of low solubility should be present in excess. Also, the temperature should be controlled at 298.16°K, the standard "room temperature." Under these standard-state conditions, the cell voltage measures ΔG°, the standard free energy change. Correspondingly, this cell-voltage is designed $\mathscr{E}^{\circ}$, the standard voltage, or as it is usually expressed, the "standard e.m.f." (e.m.f. ≡ emf ≡ "electromotive force"). Yet another term used for voltage (or emf) is "potential." In this usage, "potential" is a noun, but its meaning ties closely to the adjectival use in "potential energy."

So, at standard cell conditions,

$$\Delta G^{\circ} = -n\mathscr{E}^{\circ}\mathscr{F} \tag{6-14}$$

and since standard free energy determines the equilibrium constant, (relation 5-51)

$$\Delta G^{\circ} = -RT \log_e K = -n\mathscr{E}^{\circ}\mathscr{F}$$

or

$$\boxed{\mathscr{E}^{\circ} = \frac{RT}{n\mathscr{F}} \log_e K} \tag{6-15}$$

Of course, R and $\mathscr{F}$ must be selected in the same units: if R is in calories, $\mathscr{F}$ should be the number 23,082. It is convenient to remember the value of $RT/\mathscr{F}$ for the standard temperature 298.16°K and with the factor 2.303 inserted to convert to logarithms to the base 10.

$$\boxed{\mathscr{E}^{\circ} = \frac{0.059}{n} \log_{10} K} \tag{6-16}$$

We might illustrate the use of (6-16) to underscore its value to the chemist. Consider again the Zn–Ag electrochemical cell studied in Chapter Four. The chemistry of the cell is

$$Zn(s) + 2Ag^{+}(aq) = Zn^{+2}(aq) + 2Ag(s) \tag{6-17}$$

with equilibrium constant,

$$K = \frac{[Zn^{+2}]}{[Ag^{+}]^2} \tag{6-18}$$

It is an easy matter to set up a cell like that shown in Figure 4-12, using one-molar concentrations (standard concentrations), and to measure the voltage with zero current, as shown in

Figure 6-1(c). The experimental result is $\mathscr{E}^{\circ} = 1.56$ vclts. The equilibrium constant K can now be calculated.

$$K = 10^{n\mathscr{E}^{\circ}/0.059} = 10^{2(1.56)/0.059} \qquad (6\text{-}19)$$
$$= 10^{53}$$

and

$$\Delta G^{\circ} = -n\mathscr{E}\mathscr{F}$$
$$= -2(1.56)(23{,}082)$$
$$= -72{,}000 \text{ cal}$$
$$= -72.0 \text{ kcal} \qquad (6\text{-}20)$$

This example is a good one since it shows how the electrochemical cell gives values of ΔG° and K that would be difficult to measure by conventional analysis. In principle, we could just mix the constituents in a beaker, let them react until equilibrium is reached and then use classical chemical analysis to determine K. However, if K has the astronomical value of 10^{53}, it is implied that when the Zn^{+2} product concentration is measurab e (say, 0.1 M), the equilibrium concentration of Ag^{+} is immeasurably low. The cell gives K with easily measured concentrations, neatly sidestepping this difficulty.

(e) CHANGE OF $\mathscr{E}$ WITH CONCENTRATION

Referring back to Chapter Five, an expression like (5-46) enables us to calculate ΔG for reaction (6-17) at any set of concentrations:

$$\Delta G = \Delta G^{\circ} + RT \log_e \frac{[Zn^{+2}]}{[Ag^{+}]^2} \qquad (6\text{-}21)$$

Or, for a general reaction,

$$xA + yB + \cdots = wC + zD + \cdots \qquad (6\text{-}22)$$

therefore,

$$\Delta G = \Delta G^{\circ} + RT \log_e \frac{(C)^w(D)^z \cdots}{(A)^x(B)^y \cdots} \qquad (6\text{-}23)$$

The concentration quotient on the right-hand side has the same form as the equilibrium expression, but it involves whichever concentrations we wish to consider. We'll call this concentration coefficient Q, so that (6-23) will be more compact.

$$\Delta G = \Delta G^{\circ} + RT \log_e Q \qquad (6\text{-}24)$$

Now, substituting (6-13) and (6-14) into (6-24), we obtain

$$-n\mathscr{E}\mathscr{F} = -n\mathscr{E}^{\circ}\mathscr{F} + RT\log_e Q \tag{6-24}$$

or

$$\boxed{\mathscr{E} = \mathscr{E}^{\circ} - \frac{RT}{n\mathscr{F}}\log_e Q} \tag{6-25}$$

With $T = 298.16°\text{K}$ and inserting the factor 2.303,

$$\boxed{\mathscr{E} = \mathscr{E}^{\circ} - \frac{0.059}{n}\log_{10} Q} \tag{6-26}$$

This famous expression for the dependence of cell voltage on concentration is often called the Nernst equation. Again it is useful to see it in use. For example, what should be the measured voltage (zero current) for a cell having a Zn^{+2} ion concentration of 0.1 M and a Ag^{+} ion concentration of 0.01 M? Expression (6-26) tells us that

$$\begin{aligned}
\mathscr{E} &= \mathscr{E}^{\circ} - \frac{0.059}{n}\log_{10}\frac{[Zn^{+2}]}{[Ag^{+}]^2} \\
&= 1.56 - \frac{0.059}{2}\log_{10}\frac{(0.1)}{(0.01)^2} = 1.56 - 0.0295\log_{10}(10^{+3}) \\
&= 1.56 - 0.0295 \cdot (+3) = 1.56 - 0.0885 \\
&= 1.47 \text{ volts}
\end{aligned}$$

The Nernst equation is usually used in the reverse sense to that in the example; to calculate $\mathscr{E}_0$ from an experimental value of $\mathscr{E}$. An electrochemical cell is assembled with carefully measured concentrations. Convenience often dictates that these concentrations be other than standard conditions. The experimental value of $\mathscr{E}$ is then measured as shown in Figure 6-1. The value of Q can be calculated, and the value of $\mathscr{E}$ has already been measured, so $\mathscr{E}^{\circ}$ is the only unknown quantity in the Nernst Equation (6-25). Thus $\mathscr{E}^{\circ}$ is determined, and hence also the equilibrium constant via (6-15), and ΔG° via (6-14). Of course the temperature dependence of $\mathscr{E}^{\circ}$ can also be measured and hence, the temperature dependence of ΔG. This, in turn, can be used to establish ΔH° and ΔS°, as described in Section 5-4. Clearly, electrochemical cells have great importance in chemistry as a means of studying equilibrium and determining thermodynamic quantities.

6-2 Solution thermodynamics

Theoretical discussions are often limited to considerations of species in the gas phase. This is because the gas phase is the simplest of all states. Individual gas molecules are far enough separated to eliminate any significant forces between them. Practical chemistry, on the other hand, is definitely not limited to gas-phase reactions. In fact, most of the chemistry we see around us occurs in solution, and most of that in the especially abundant solvent, water. That is why the study of equilibrium in ionic aqueous solutions forms a major part of any introductory chemistry course. In this section we will discuss the thermodynamics of some simple ionic, aqueous solutions, paying special attention to the extremely important role played by the solvent, water. Table 6-1 lists some thermodynamic properties of ions in aqueous solution and, for comparison, some gas-phase values as well.

Table 6-1 Thermodynamic Properties of Selected Gaseous and Aqueous Ions and Their Compounds

	ΔH_f° (kcal/mole)	ΔG_f° (kcal/mole)	S° (cal/mole degK)
H^+ (aq)	0	0	0
(g)	+367.2		
Na^+ (aq)	−57.4	−62.6	14.1
(g)	+146.0		
Ca^{++}(aq)	−129.8	−132.2	−13.2*
Ag^+(aq)	25.2	18.4	17.4
Cl^-(aq)	−39.9	−31.4	13.5
(g)	−58.8		
OH^-(aq)	−54.9		
(g)	−33.7		
F^-(aq)	−79.5	−66.6	−3.3
HF(g)	−64.2	−64.7	+41.5
HCl(g)	−22.1	−22.8	+44.6
NaOH(s)	−101.7		
CaF_2(s)	−290.3	−277.7	16.46
AgCl(s)	−30.4	−26.2	23.0
NaCl(s)	−98.2	−91.8	17.3

*Negative entropies occur because of the choice of $S^\circ[H^+(aq)] = 0$.

The solution values are referred to the properties of H^+(aq) as a reference state. Accordingly, *the thermodynamic properties of H^+(aq) are defined to be zero.* Remember that we are always concerned with differences, so that our choice of a reference point does not matter (recall the building contractor in Denver).

(b) ENERGY EFFECTS IN IONIC SOLUTIONS

Hydrochloric acid and sodium hydroxide are among the most familiar chemical reagents. Each of them dissolves readily in water to form an ionic solution that apparently contains no undissociated molecules. That is, the processes

$$HCl(aq) \rightarrow H^+(aq) + Cl^-(aq) \tag{6-27}$$

and

$$NaOH(aq) \rightarrow Na^+(aq) + OH^-(aq) \tag{6-28}$$

go effectively to completion.

Anyone who has ever dissolved sodium hydroxide in water or bubbled HCl gas into water knows that each solution is warmed perceptibly. This amateur calorimetry tells us that both these processes are exothermic. But doesn't it take energy to separate these molecules into ions? Let us examine the energetics of the *gas-phase* reactions using the data in Table 6-1.

$$HCl(g) \rightarrow H^+(g) + Cl^-(g)$$
$$\Delta H^\circ = +330.5 \text{ kcal/mole} \tag{6-29}$$

$$NaOH(s) \rightarrow Na^+(g) + OH^-(g)$$
$$\Delta H^\circ = +214.0 \text{ kcal/mole} \tag{6-30}$$

We have considered the dissociations beginning from the states in which HCl and NaOH are found at room temperature. Both these reactions absorb an enormous amount of heat—they are highly endothermic. Thermodynamic data for the entropies associated with these reactions are not available, but even though we expect an entropy increase, it could not possibly be large enough to prevent ΔG from being positive. We can be confident, then, that neither of these gas-phase reactions will occur spontaneously.

The story is quite different in aqueous solution. We have already observed that these substances dissolve with the *evolution* of heat. Let's use Table 6-1 to find out how much.

$$HCl(g) \rightarrow H^+(aq) + Cl^-(aq)$$
$$\begin{aligned}\Delta H^\circ &= \Delta H_f^\circ(H^+) + \Delta H_f^\circ(Cl^-) - \Delta H_f^\circ(HCl)\\ &= 0 + (-39.9) - (-22.1)\\ &= -17.8 \text{ kcal/mole}\end{aligned} \tag{6-31}$$

and

$$NaOH(s) = Na^+(aq) + OH^-(aq)$$
$$\begin{aligned}\Delta H^\circ &= \Delta H_f^\circ(Na^+) + \Delta H_f^\circ(OH^-) - \Delta H_f^\circ(NaOH)\\ &= (-57.4) + (-54.9) - (-101.7)\\ &= -10.6 \text{ kcal/mole}\end{aligned} \tag{6-32}$$

How is it possible that identical processes carried out in the gas phase and in solution have such different energy changes? It

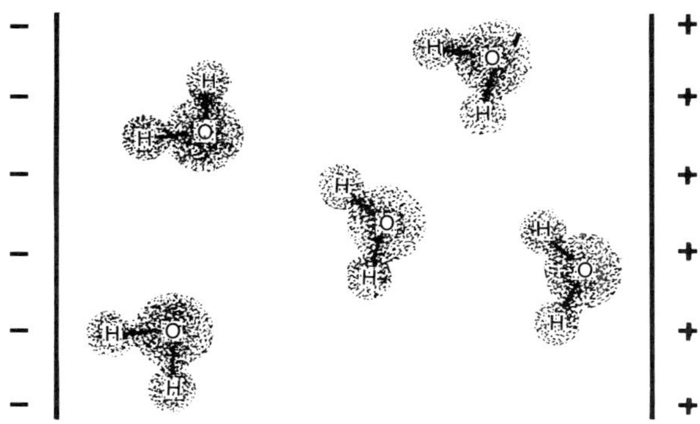

Figure 6-2 Water molecules placed between electrically charged plates orient themselves with the oxygen end pointing towards the positive plate.

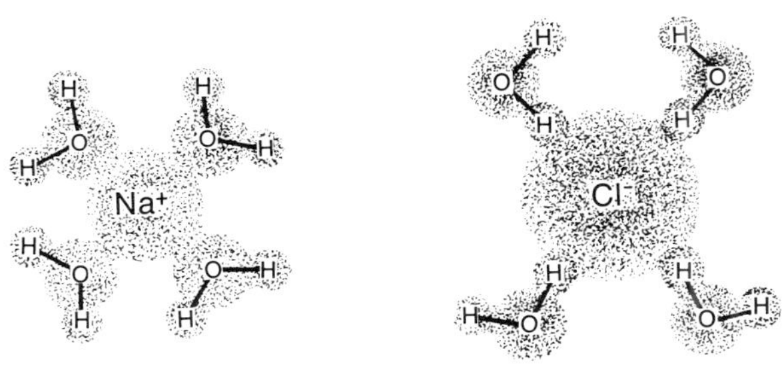

Figure 6-3 $Na^+(aq)$ and $Cl^-(aq)$ are stabilized by association with polar H_2O molecules.

is clear that the solvent must play the crucial role. Indeed reactions (6-31) and (6-32) are much more complex processes than indicated by the equations. Each ion is surrounded by a number of water molecules, forming a complex ion in solution. This results because water molecules have an electric dipole moment: the hydrogens have a net positive charge and the oxygens a net negative charge (see Fig. 6-2). Dissociation of HCl and NaOH into ions is possible because the ions are stabilized in solution by the polar water molecules. Schematic diagrams of the environment of sodium and chlorine ions are shown in Figure 6-3. In the case of $H^+(aq)$, a stable species $H_9O_4^+$ is formed, which is generally considered to be an H_3O^+ molecule stabilized by three more water molecules, as shown in Figure 6-4. Thus, the ionization of HCl is not just the separation of a positively charged proton from

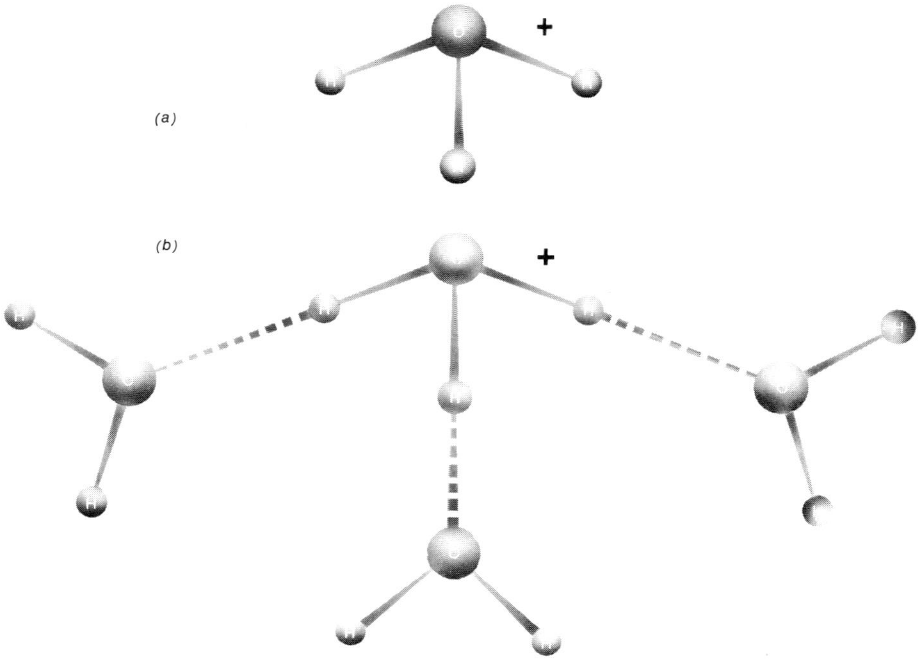

Figure 6-4 A possible structure of $H_9O_4^+$ based upon H_3O^+.

a negatively charged chloride ion, as in the gas phase. Instead, the proton is removed from the neighborhood of a chloride ion, and both ions are placed in comparably favorable neighborhoods formed by the dipolar solvent molecules in the species we designate $H^+(aq) \equiv H_9O_4^+$ and $Cl^-(aq)$.

Thus, we find that strong acids and bases are stabilized in solution by the formation of complexes with the solvent water. This effect lowers the energy by a very large amount—enough, in fact, to overcome the extremely high energy needed to carry out the same processes in the gas phase.

(b) THERMODYNAMICS OF SOLUBILITY

Chemical substances known as "salts" are strongly bonded, ionic solids. Most salts that dissolve in water are totally dissociated into ions. The variation of solubility encountered in different salts is high, however. What factors affect the solubility? Well, from the previous discussion we can expect the solvent to play a very large role. Consider the two salts, sodium chloride and calcium fluoride. Considerable energy is needed for the dissociation into ions in the gas phase, yet both dissolve in water with the absorption of only a small amount of heat.

$$NaCl(s) \rightarrow Na^+(aq) + Cl^-(aq) \qquad (6\text{-}33)$$

$$\Delta H^\circ = +0.9 \text{ kcal/mole}$$

$$CaF_2(s) \rightarrow Ca^{++}(aq) + 2F^-(aq) \qquad (6\text{-}34)$$
$$\Delta H^\circ = +1.5 \text{ kcal/mole}$$

The experimental solubilities of these two salts are very different: sodium chloride dissolves to a concentration above 5 moles per liter, while calcium fluoride has an equilibrium solubility below 0.001 moles per liter. Entropy changes associated with the two processes must account for the difference, since the enthalpy changes are nearly identical. We would probably predict that ΔS° would be greater in the CaF_2 dissociation, since three ions are being produced rather than two, as in the case of NaCl. This would cause the CaF_2 solubility to be the higher of the two. In fact, just the opposite is true. Sodium chloride is quite soluble, whereas calcium fluoride is barely soluble at all.

Turning to the thermodynamic properties summarized in Table 6-2, we see that the entropy of CaF_2 does not meet our expectations. Not only is the entropy change less for CaF_2 than for NaCl; it is actually negative! The aqueous, ionized state is less random than the solid. This points out dramatically the important effect of the solvent on solution processes. The small fluoride ions and highly charged Ca^{+2} ions are tightly surrounded by several water molecules, in arrangements that are more ordered than their environments in the crystal! Apparently the larger size of the chloride ions and the lower charge of the sodium ions makes these effects much less pronounced for NaCl.

Table 6-2 Thermodynamics of Salt Solubilities in Water (298°K)

	ΔH° (kcal/mole)	ΔS° (cal/mole degK)	ΔG° (kcal/mole)
$NaCl(s) = Na^+(aq) + Cl^-(aq)$	+0.9	+10.3	−2.1
$CaF_2(s) = Ca^{+2}(aq) + 2F^-(aq)$	+1.5	−36.3	+12.3

Similar results are found for gas solubilities. Hydrogen chloride gas is roughly one thousand times more soluble in water than is hydrogen fluoride gas. The thermodynamics of these processes is given in Table 6-3. We see that not only is HF less soluble in

Table 6-3 Thermodynamics of Gaseous Solubility (298°K)

	ΔH° (kcal/mole)	ΔS° (cal/mole degK)	ΔG° (kcal/mole)	K
$HF(g) = HF(aq)$	−11.7	−20.3	−4.6	$2.3 \cdot 10^3$
$HF(aq) = H^+(aq) + F^-(aq)$	−4.0	−23.5	+3.2	$4.5 \cdot 10^{-3}$
$HF(g) = H^+(aq) + F^-(aq)$	−15.7	−43.8	−1.4	10.4
$HCl(g) = H^+(aq) + Cl^-(aq)$	−17.8	−31.1	−8.6	$2.0 \cdot 10^6$

water than HCl, but it is also not significantly dissociated into ions! The reason is apparent in the thermodynamics. After dissolving in water, the HF would decrease the solution entropy by another 23.5 cal/degK if it were to dissociate into ions. Again, the small size of the F^- ion causes it to arrange solvent molecules strongly around itself to produce a highly ordered solution. So entropy effects cause the behavior of aqueous HF and aqueous HCl to be entirely dissimilar. Hydrofluoric acid is a weak acid with only moderate solubility in water, whereas hydrochloric acid is a very strong acid that dissolves to concentrations above 10 moles per liter.

(c) REACTIONS IN SOLUTION

Solvent effects have an equally important influence on the outcome of chemical reactions occurring in solution. One of the crucial decisions in bringing about a reaction is the choice of a suitable solvent. Water, our most abundant chemical, is frequently our choice, particularly when the reactions involve ions.

Table 6-4 Thermodynamics of Ionic Reactions (298°K)

	ΔH° (kcal/mole)	ΔS° (cal/mole degK)	ΔG° (kcal/mole)	K
(i) $HF(aq) + F^-(aq) = HF_2^-(aq)$	+ 0.7	+ 4.2	− 0.7	3.3
(ii) $HOCl(aq) + OCl^-(aq)$ $= H^+(aq) + Cl^-(aq) + ClO_2^-(aq)$	− 1.3	− 6.3	+ 9.6	9.1×10^{-8}
(iii) $NH_3(aq) + HF(aq) = NH_4^+(aq) + F^-(aq)$	−14.8	−24.0	− 8.35	1.4×10^{6}
(iv) $OH^-(aq) + H_3O^+(aq) = 2H_2O$	−13.3	+19.3	−19.1	1.0×10^{14}

Two pairs of reactions are contrasted in Table 6-4. Reaction (i), an endothermic reaction, results in the formation of one product species from two reactant ones. By itself, this should make the system less random (two going to one). In fact, the expected decrease in entropy does not occur. Quite the opposite, the entropy *increases* enough, in fact, to overcome the endothermicity and produce a negative free energy change. The reason for this effect is, again, the highly ordered hydration around the small fluoride ion. This order is lost (the system becomes more random) as the larger HF_2^- ion is formed. The resulting increase in randomness more than overcomes the fact that two particles combine to form one.

Reaction (ii) shows the opposite effect. In this reaction, two reactant molecules give three product species—randomness should increase. Instead, ΔS is negative! Again the effect is due to the production of three ions, each of which creates an ordered hydration environment around itself. This means an entropy decrease despite the production of three molecules from two. Now negative entropy overpowers an exothermic ΔH to make ΔG positive and the equilibrium constant very small.

These effects are strikingly underscored by reactions (iii) and (iv) in Table 6-4. These processes involve identical numbers of product and reactant species and are exothermic by nearly the same amount. However reaction (iii), which involves the formation of ions from neutral species, has a negative ΔS, while reaction (iv), in which ions are converted into neutral water molecules, has a positive ΔS. The difference, entirely due to positional randomness, causes a factor of 10^8 difference in the equilibrium constants!

Thus we see that aqueous solution chemistry cannot be predicted on the basis of enthalpy considerations alone. The hydration entropy of ions is often important, and sometimes dominant, in determining the reactions that occur.

6-3 Some things about entropy

The entropy of a mole of molecules contributes importantly to the reactions of those molecules. What factors determine the entropy of those particular molecules? According to Chapter Three, it is connected with their positional and motional randomness at the temperature in question. The positional randomness depends upon the state: gas, liquid, solution, or solid. The motional randomness depends upon the number of degrees of freedom: translational, rotational, and vibrational—and their energy spacing. We can learn much about the entropy of a substance by merely considering how all of these degrees of freedom are affected as the substance is cooled, first as a gas, then, after condensation, as a liquid, and finally, as a solid.

(a) TEMPERATURE AND RANDOMNESS

The average translational energy of a gaseous molecule is $(3/2)kT$. Hence, as the temperature drops, the energy to be divided among the translational states is less and less. Thinking of the energy as being in packets, we have fewer and fewer energy packets to distribute among the molecules. Their motional randomness, hence their entropy, drops. The same qualitative effect of lowered temperature is felt among the rotational and vibrational degrees of freedom. Fewer and fewer states have substantial occupancy, so randomness drops. *The entropy of any gas drops as it is cooled.*

At some temperature the gas will liquefy. Then the translational motion of the molecules is very much restricted, and positional randomness is drastically curtailed. The molecules still move about, randomly vibrating and tumbling against each other, but without anything like the positional freedom of the gas phase. Randomness drops precipitously as the gas changes to the liquid state. *The entropy of any gas drops sharply when it liquefies.*

Cooling the liquid below the condensation temperature lowers still further the amount of molecular motion. Again fewer and fewer energy states are occupied—motional randomness is decreased as temperature continues to fall. *The entropy of any liquid drops as it is cooled.*

When the freezing point is reached, molecules settle into the regular positions of the crystal lattice, sacrificing the orientational disorder of the liquid state and more of the motional freedom. To be sure, the crystal has its own vibrational modes, but these reflect the rigidity and regularity of the solid state. *The entropy of any liquid drops sharply when it solidifies.*

Finally, what happens as we continue to cool the solid? As the temperature drops still further, there is less and less vibrational movement of the molecules about their equilibrium positions. Positional randomness is now virtually nonexistent, and motional randomness is disappearing. In fact, as absolute zero is approached, there is less and less energy available to excite any kind of molecular movement. It is intuitively reasonable that all random motion should stop at absolute zero. Then every molecule is fixed in an equilibrium lattice site, and there are no random vibrations excited. This describes a system without any randomness at all! That means the entropy, which measures disorder, has reached zero! Thinking of the relationship of entropy to probability, $S = R \log_e W$, at absolute zero, there is only one arrangement, that of the perfect crystal, so $W = 1$ and $S = R \log_e 1 = 0$. *At the absolute zero of temperature, the entropy of all substances becomes zero.* This generalization can be taken as an experimental fact as well as an intuitively reasonable proposal. The experimental evidence is so complete and so convincing, that this statement is known as the Third Law of Thermodynamics.

The Third Law of Thermodynamics:
The entropy of any substance is zero at absolute zero temperature. *(6-29)*

(b) THE THIRD LAW AND ABSOLUTE ENTROPIES

With this universal reference point, we can specify entropies on an absolute scale. Each molecular type has a natural reference state, so we need no arbitrary choice. Hence tabulated entropies are specified as "absolute entropies." The entropy is simply measured by heating the solid from the lowest accessible temperature up to the temperature and state of interest, always keeping track of the quantity q_{rev}/T. Figure 6-5 shows a typical experimental entropy plot, measured from 4°K up to room temperature. The two sharp rises are connected with melting and vaporization.

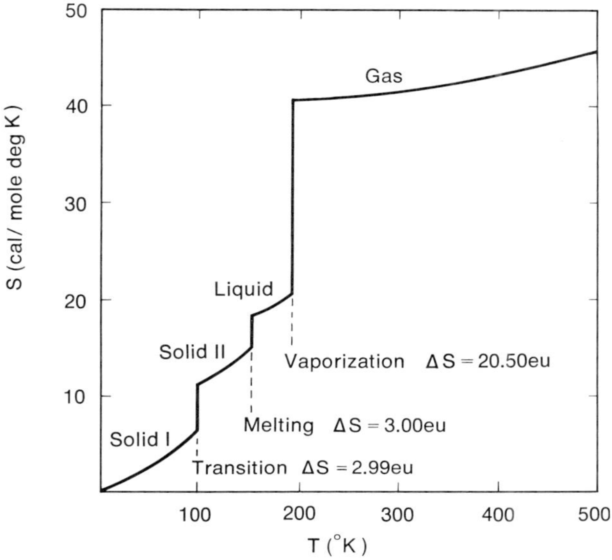

Figure 6-5 Experimental measurement of the entropy of SO_3 as a function of temperature.

These changes are characteristic, as mentioned in Section (a), and as shown for a number of substances in Table 6-5.

Table 6-5 Dependence of Entropy on State (cal/mole degK)

Substance	S^0_{298}(s)	$S^0_{298}(\ell)$	S^0_{298}(g)
SO_3	12.5	22.9	61.3
S	7.6		40.1
I_2	27.8		62.3
Br_2		36.4	58.6
BrF_3		42.6	69.9
CH_3OH		30.3	57.3

(c) ENTROPY AND SIZE

A monatomic gas has no vibrations or rotations, so all its energy randomness must come from its translational energy. The kinetic theory tells us that the number of ways in which a particle can hold a given energy of translation depends on the mass of the particle. In our energy-packet description, the size of the energy packets gets smaller as the particle mass increases. This implies that a given amount of translational energy creates greater randomness when distributed among a mole of heavy particles than when distributed among a mole of light particles. Since entropy depends on the logarithm of the probability (see Section 4-3), a plot of entropy against the logarithm of atomic weight should be linear for simple atomic gases. This expectation is borne out by the data in Table 6-6, which are plotted in Figure 6-6.

Table 6-6 Entropies of Monatomic Gases

Species	Atomic Weight	Entropy (cal/mole degK)
He	4.00	30.12
Ne	20.2	34.95
Ar	40.0	36.98
Kr	83.8	39.2
Xe	131.3	40.5

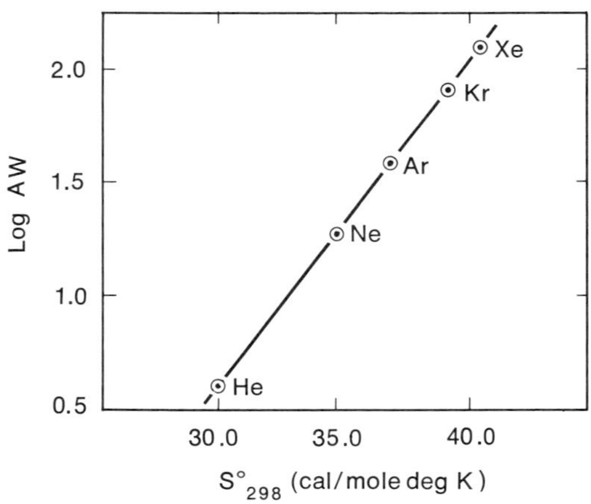

Figure 6-6 Entropies of monatomic gases.

This weight dependence is found if we compare molecules of similar shape but different size. Figure 6-7 shows plots of log(MW) against S°_{298} for simple hydrides containing from two to five atoms. In every case the entropy increases linearly with molecular weight.

(d) ENTROPY AND MOLECULAR COMPLEXITY

The intrinsic energy randomness of a substance is related to the ways in which vibration and rotation, as well as translation, can soak up energy. Table 6-7 and Figure 6-8 show how the absolute entropy changes with increasing molecular complexity for a more or less fixed molecular weight. There is seen to be a general increase as the number of atoms increases. The more atoms there

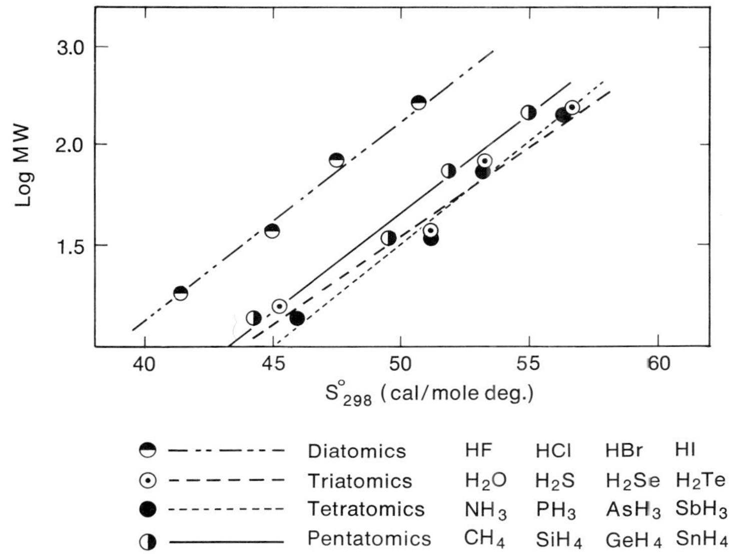

Figure 6-7 Variation in S^0_{298} with log(MW) for simple hydrides.

Table 6-7 Entropies of Gases as a Function of Increasing Molecular Complexity for Fixed Molecular Weight

Molecule	Number of Atoms	Molecular Weight	S^0_{298}	Range M.W.
Ne	1	20.18	34.95	
HF	2	20.01	41.51	
D_2O	3	20.02	47.74	19–20
ND_3	4	20.03	49.13	(~20)
CHD_3	5	19.05	49.98	
Ar	1	39.95	36.98	
F_2	2	38.00	48.44	
CO_2	3	44.01	51.06	34–44
H_2O_2	4	34.01	55.6	(~40)
CH_2N_2	5	42.04	58.02	
Kr	1	83.80	39.19	
Cl_2	2	70.91	53.29	
CS_2	3	76.14	56.82	70–85
SO_3	4	80.06	61.34	(~80)
CH_2Cl_2	5	84.93	64.56	
Xe	1	131.30	40.53	
BrCl	2	115.36	57.36	
NOBr	3	109.92	65.38	98–131
$COCl_2$	4	98.92	67.74	(~120)
$CHCl_3$	5	119.38	70.65	

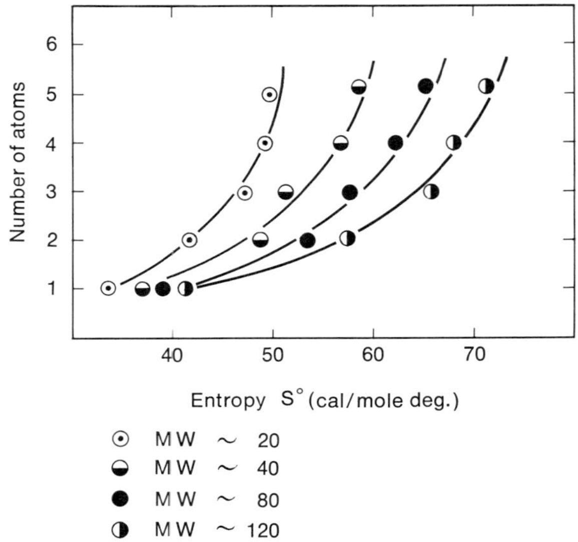

Figure 6-8 Entropy variation with number of atoms for fixed molecular weight.

are in the molecule, the more ways there are in which it can vibrate.

6-4 Thermodynamics in Living Systems: Bioenergetics

Every spring we witness a seeming thermodynamic miracle. From the brown, raw earth, wet with rain and warmed by the sun, there rises a vast family of identical sprouts. Then as summer comes, the various chemical mixtures in the soil are transformed into fully developed blades of grass. Each blade is like every other, a beautifully organized, highly efficient chemical factory. The whole process takes place spontaneously—in the direction of *increasing* randomness! The starting materials looked random enough—the wet earth with its complex chemical mixture—the final result, the blade of grass, seems far from random! Thermodynamics, wherefore art thou?

The Second Law of Thermodynamics speaks, of course, in terms of the entropy of the universe. True, the processes by which living systems grow and reproduce succeed in creating locally a high degree of order; that is, within the blade of grass, within the cell, within the DNA molecule. But how do these processes drain the surroundings? There we must find the "driving force" to support the local creation of order. There we must find a counterbalancing increase in randomness to make the processes of life occur spontaneously. There we must find our thermodynamic principles at work, making it possible for randomness to *decrease* locally, at the cost of an even larger increase in randomness elsewhere!

Here on Earth, the wellspring of life is the Sun. As this giant inferno consumes itself, wasting itself for all time, the tiniest fraction of its ever increasing randomness more than suffices to create a garden of order—life—on our lush planet. It does so by sending us light, a form of energy, and we must take it from there.

(a) ENERGY FROM THE SUN

Man fuels his body with both plant and animal tissue. However, animals consumed as food, themselves rely mainly on plant life for their nourishment. The plant world is thus the source of all food and energy for the animal world. Ultimately all energy has its origins in the fiery furnace that is the Sun. There, extremely high temperatures result from the release of the energies associated with the attractions between nuclear particles. The principal process going on is the "fusion" of hydrogen atoms to form helium, a process that releases enormous quantities of energy in the form of heat.

$$4\text{H} \rightarrow \text{He} + 2 \text{ electrons} \quad \Delta H = -5.7 \cdot 10^{8} \text{ kcal/mole He}$$

This is an endothermic reaction that really tries! Where such a fire burns, temperatures reach millions of degrees (near the center of the Sun). Even on the cooler outer edges, the solar gases swirl at 6000°K temperatures. This is the part of the Sun we see; the inner tempest is shrouded from us, fortunately, or the Earth would have a brown, cooked surface and life could not exist.

So we see these outer gases at 6000°K, a warm enough climate to break molecules down to atoms and to strip electrons off the atoms. This hot, charged cloud emits light of blinding intensity. The light pours out into space, and far away the planets intercept only a minute fraction of it. The rest goes on to the stars, who see our Sun as another tiny, insignificant dot in their heavens, a dot so dim that it could hardly be interesting.

Here on Earth, however, the atmosphere, the soil, and the oceans are warmed to produce a reasonably well thermostatted environment with a mean temperature such that molecules can exist but also undergo chemical transformations at a rate well tuned to the day–night time cycle. In this favorable setting, chlorophyll in green plants also absorbs sunlight and uses the energy for the synthesis of a sugar, glucose, from water and atmospheric carbon dioxide. This process, illustrated in Figure 6-9, is the beginning of the biological processes that are responsible for the manufacture of all living matter, from plankton to planaria to people.

It is clear from ΔH° and ΔG° that the synthesis of glucose is an uphill fight. Energy has to be supplied. It is estimated that over

35×10^9 tons of carbon per year are made into glucose by photosynthesis, in land and marine plants. This requires an estimated 10^{18} kilocalories per year of energy from the Sun. Even this enormous amount of energy is only a drop from the bucket of total

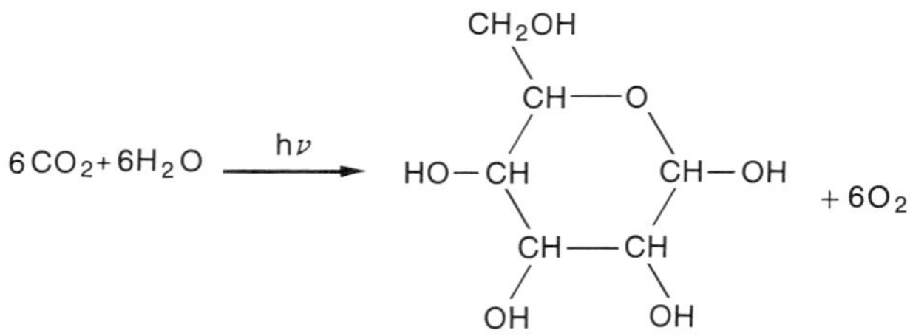

Glucose, $C_6H_{12}O_6$

$\Delta H° = +673$ kcal/mole

$\Delta G° = +686$ kcal/mole

energy received on the surface of the Earth—in excess of 10^{21} kilocalories per year. Yet the total amount of energy expended in a given year by all manmade machines is guessed to be only about 10^{16} kilocalories. So in using up one one-thousandth of the light energy from the Sun, the plant factories channel into useful purposes one hundred times more energy than that which is used by man with all his machines. It is clear then, why the study of bioenergetics is of great importance.

(b) BIOLOGICAL ENERGY TRANSFER

Energy originating in hydrogen fusion on the surface of the sun is absorbed by green plants and stored as chemical potential energy in the glucose molecule. How is this energy utilized to build cells or contract muscles in living things? At the heart of the biological system of energy transfer is a remarkable pair of compounds, adenosine diphosphate (ADP) and adenosine triphosphate (ATP). These molecules are rather complex organic ring molecules, but we are not concerned with the details of their structure. The key to nearly all movement of energy in biological systems is the following reaction, which represents the hydrolysis of adenosine triphosphate by water to produce adenosine diphosphate and a hydrogen phosphate ion.

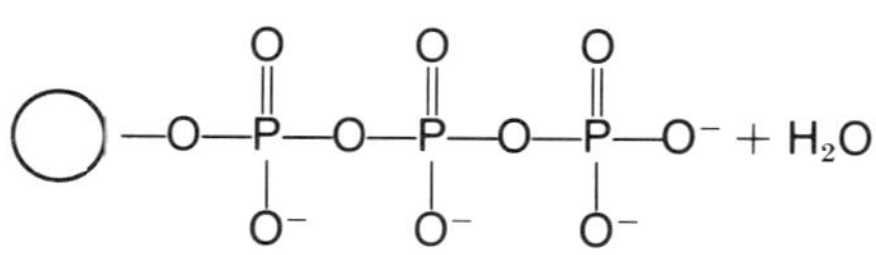

$(ATP)^{-4}$

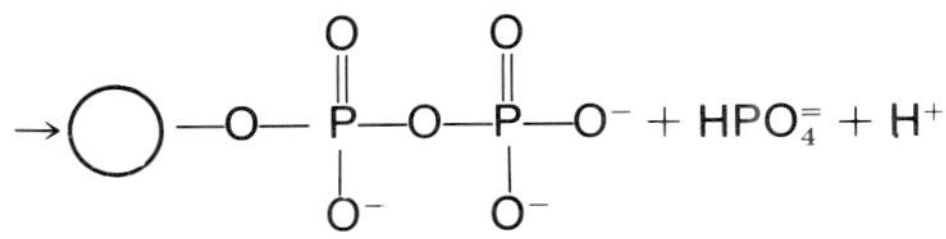

$(ADP)^{-3}$ *(6-30)*

The adenosine fragment is represented schematically by a circle. The free energy change associated with this reaction is

$$\Delta G^{o}(\text{ATP} \rightarrow \text{ADP}) = -7 \text{ kcal/mole} \qquad (6\text{-}31)$$

The free energy change is negative, so the reaction can proceed spontaneously. The hydrolysis of a phosphorus–oxygen bond can be used to perform a maximum of 7 kilocalories per mole of chemical work. This process is responsible for nearly all energy transfer and order-building processes in living things. We shall now see how this is performed.

(c) ATP ENERGY FLOW

A flow diagram for biological energy is shown in Figure 6-9.

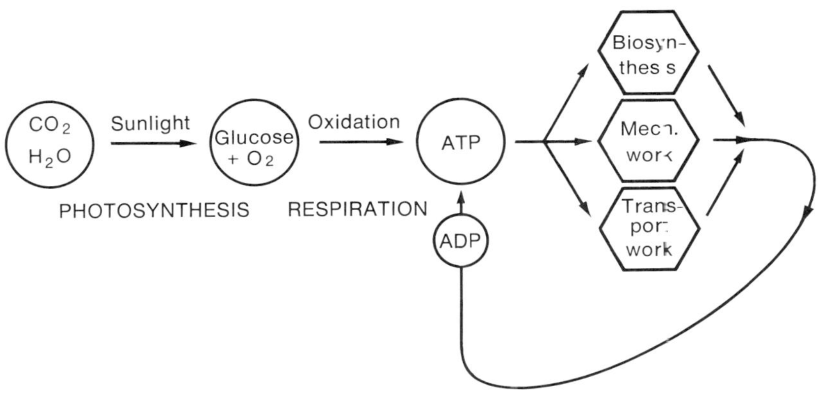

Figure 6-9 Flow of biological energy. Energy absorbed during photosynthesis is transferred to ATP in the respiration step. ATP energy is employed to perform biological work—the resulting ADP then begins the cycle again.

Energy stored in glucose is released by the process of oxidation, which is just the reverse of the synthetic process resulting from photosynthesis.

$$\text{glucose} + 6O_2 \rightarrow 6CO_2 + 6H_2O \qquad (6\text{-}31a)$$
$$\Delta H^{o} = -673 \text{ kcal/mole}$$
$$\Delta G^{o} = -686 \text{ kcal/mole}$$

This step is called *respiration.* Energy released during respiration is used to form ATP from ADP (the reverse of (6-30)). The ATP energy is then employed to do useful work in the cells. This work is commonly divided into three categories: biosynthesis (building of new cells), mechanical work (muscle contraction) and transport work (movement of fluid between regions of differing concentrations). The free energy available on hydrolysis of the P—O bond in ATP is utilized by means of "coupled reactions," some of which we will now describe.

(d) ATP FORMATION

The oxidation of sugars back to carbon dioxide and water takes place in over seventy separate steps, each one catalyzed by a separate enzyme. One such step might be the oxidation of an aldehyde, RCHO to a carboxylic acid, RCOOH.* In aqueous solution this process proceeds with a large decline in free energy.

$$\underset{\text{aldehyde}}{\mathrm{RCHO}} + \tfrac{1}{2}\mathrm{O_2} \rightarrow \underset{\text{acid}}{\mathrm{RCOO^-(aq)}} + \mathrm{H^+(aq)} \quad \Delta G = -7\ \text{kcal/mole} \qquad (6\text{-}32)$$

On the other hand, the production of ATP from ADP and phosphate is uphill on the free energy surface by about this same amount. Clearly, a combination of these two processes would have a ΔG near zero and would result in the formation of ATP without an accompanying increase in free energy. A sequence of steps that brings about this net result is possible. Such a process is known as a coupled reaction.

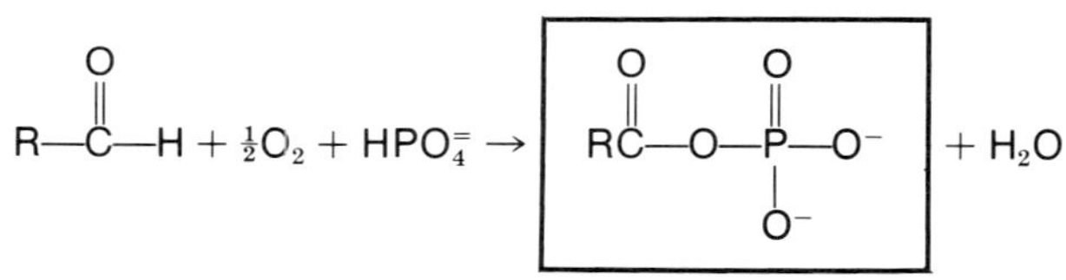

(6-33a)

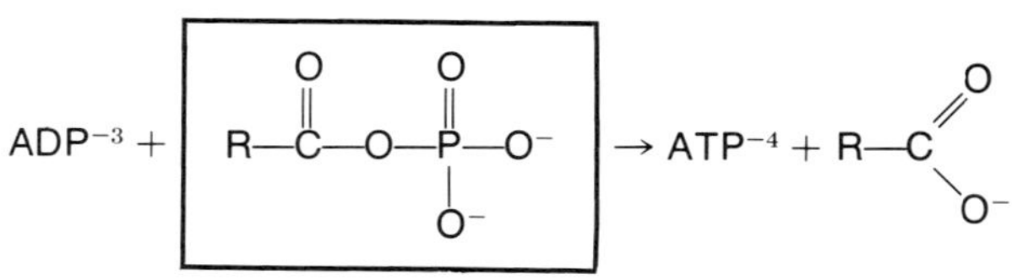

(6-33b)

Each of these steps must be catalyzed by an appropriate enzyme. The large free energy change associated with the aldehyde oxidation (6-33a) is conserved by the formation of the phosphate ester (box). The free energy of hydrolysis of a phosphate ester is even larger than that of ATP itself. Thus, when the second reaction occurs, the enzymatic transfer of a phosphate group from the ester to ADP, this free energy is used for

*We will be utterly unspecific about the organic compounds we mention. The symbol R is used to represent some organic fragments. For example, if we were considering the oxidation of acetaldehyde to acetic acid, $CH_3CHO \rightarrow CH_3COOH$, R would represent the methyl group CH_3.

the formation of ATP. Now the favorable free energy change associated with the oxidation is effectively conserved as ATP free energy, which is, in turn, used in further reactions. This is possible only because the pair of reactions (6-33) have a common intermediate, the phosphate ester.

(e) ATP AT WORK IN THE SYNTHESIS OF SUCROSE

The sugar sucrose (common cane sugar) is made from glucose and fructose (another sugar, very similar in structure to glucose). The free energy change for this process is +5.5 kilocalories per mole. Free energy will not permit the reaction to proceed spontaneously without some help. Biologically it is carried out in a sequence of reactions in which a glucose phosphate complex is formed by reaction with ATP. We can calculate the efficiency of this process, and write a suitable route for the reaction.

The net reaction is

$$\text{glucose} + \text{fructose} + \text{ATP} \rightarrow \text{sucrose} + \text{ADP} \qquad (6\text{-}34)$$

Consider the two component reactions

$$\text{glucose} + \text{fructose} \rightarrow \text{sucrose} \quad \Delta G = +5.5 \text{ kcal/mole} \qquad (6\text{-}35)$$

$$\text{ATP} \rightarrow \text{ADP} + \text{phosphate} \quad \Delta G = -7.0 \text{ kcal/mole} \qquad (6\text{-}36)$$

If the performance of 7 kilocalories per mole of work by the ATP/ADP conversion results in only 5.5 kilocalories per mole of work in the synthetic sequence, the overall process is

$$\frac{5.5}{7.0} \times 100 = 79\% \text{ efficient}$$

A full sequence of steps, in which the original free energy comes from aldehyde oxidation would be:

$$\text{RCHO} + \tfrac{1}{2}\text{O}_2 + \text{HPO}_4^= \rightarrow \boxed{\text{RCOOPO}_3^=} + \text{H}_2\text{O}$$

$$\boxed{\text{RCOOPO}_3^=} + \text{ADP}^{-3} \rightarrow \boxed{\text{ATP}^{-4}} + \text{RCOO}^-$$

$$\boxed{\text{ATP}^{-4}} + \text{glucose} \rightarrow \text{ADP}^{-3} + \boxed{\text{glucose-phosphate}}$$

$$\text{fructose} + \boxed{\text{glucose-phosphate}} \rightarrow \text{sucrose} + \boxed{\text{HPO}_4^=}$$

$$\text{RCOH} + \tfrac{1}{2}\text{O}_2 + \boxed{\text{HPO}_4^=} \rightarrow \cdots\cdots$$

In each step, the free energy of the RCHO is transferred from intermediate to intermediate, permitting the sequence of reactions to occur. This stepwise utilization of the available free energy is suggested by the boxes around the intermediates in the sequence.

(f) BIOLOGICAL INFORMATION AND ENTROPY

A child playing with a typewriter could eventually fill a page with letters. To others than, possibly, himself, this page would be totally meaningless—a random collection of letters and spaces. On the other hand, consider the decisions made in writing a page of this book.* Information transmitted via the printed word is a culmination of an enormous number of decisions: after each letter, the author has to choose, from among the twenty-six letters of the alphabet, the next one which will create the word which, when assembled with other words, conveys his idea.

We have already discussed energy and positional randomness. Let's consider a new kind, *information randomness.* The two pages described above might contain exactly the same assortment of letters. On the child's page, the information randomness is very high; on the second the randomness has decreased in favor of the information contained in the ordered sets of words. The ordering of the letters in the alphabet into words, and the words into sentences, represents a very large decrease in informational randomness. In the language of thermodynamics we would say that the transformation from the child's game to the manuscript page involved a very large decrease in entropy. For this transformation to occur, other processes must occur in which entropy of the surroundings increases even more. As these processes proceed down the free energy hill, a part of the free energy is stored in the informational content of the page. Consideration of information as stored free energy (or stored "order") is a part of the new field "information theory," an extension of both thermodynamics and probability theory. One of the most interesting applications of information theory is to biological systems. We will not go into detail about the methods, but describe very briefly one of the ways in which biological information is stored.

Biological instructions for the manufacture of cells in living matter are stored in chemical substances. The most well known example is DNA (deoxyribonucleic acid), a macromolecule apparently responsible for transmitting hereditary characteristics. The "alphabet" used by DNA to record biological instructions is short, consisting of only four "letters." These letters are really complex molecules, attached in a specific way to the very long-chain DNA molecule. The ordering of these four molecules on DNA present in genes tells the biosynthetic factories whether

*Any suggestion of similarities between the two situations here described would be considered unkind by the authors. . . .

they should be producing blond or brunette, fingers or toes, four legs or two, and so on.

The amount of biological information stored in even a very simple cell is enormous. When put on a quantitative basis by information theory, it is found that even a very small cell contains more information than a volume of an encyclopedia. From a probability point of view, then, the odds against the processes that store the information are overwhelming. It is only because the entropy decrease in these processes is balanced by the entropy increases associated with the formation of small molecules (CO_2, H_2O, etc.) as by-products of the combustion processes, that life is maintained.

In conclusion, then, we have seen that all biological processes ultimately depend upon energy from the sun to build complex molecules. These complex molecules store both energy and order, the former to activate the biological system, and the latter to provide a downhill, free energy path on the way. Finally, the ultimate key to reproduction, chemical information storage, is accomplished through molecular replication. This process is probably the most entropy-expensive step in the chemistry of life—and the most essential.

appendix a
integration

In scientific problems it often becomes necessary to measure the area under a curve. We first met this in Section 2-2(f) when discussing pressure–volume work. We found that the work done by an expanding gas can easily be calculated if the pressure remains constant

$$w(P \text{ constant}) = P\Delta V$$

Graphically, the work is numerically equivalent to the area under a plot of P against V between the initial volume V_1 and the final volume V_2 as illustrated in Figure 2-5*a.* We want to examine now what happens if, as in Figure 2-5*b* and Figure A-1, the gas pressure does not remain constant, but changes in some regular way during the expansion. Is the area under the curve numerically equal to the work done in this case also? We can show that it is by the following reasoning.

Let's begin by drawing a rectangular path from the initial state P_oV_o to the final state P_fV_f that will let us calculate the work done. This path, shown on Figure A-2, involves an increase in pressure at constant volume during which $w=0$ since $\Delta V=0$. The second step is a constant pressure expansion with the work given by

$$w = P_f\,(V_f - V_o) = P_f\,\Delta V$$

Clearly this is not the correct answer. We have done our calculation on a path much different from the desired one. We can come closer to the real path by the process illustrated in Figure A-3. Here we have divided the volume change into two equal parts, $\Delta V'$, and drawn constant pressure lines to the final pressure points in both halves. Now the work done is given by a sum of that done in each of the two steps

$$w = P_1\,\Delta V' + P_f\,\Delta V'$$

We still do not have a path that agrees with the desired one, but obviously we have come closer.

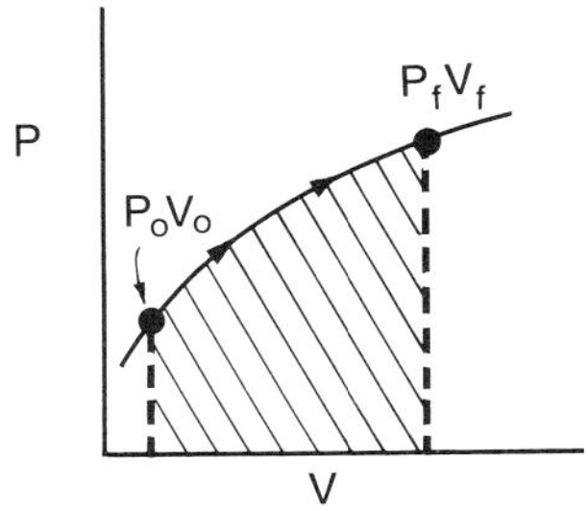

Figure A-1

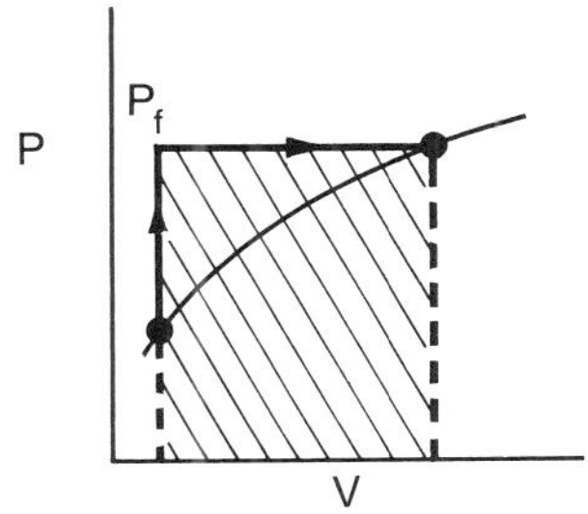

Figure A-2

If we continue this process, taking smaller and smaller volume intervals, we will come closer and closer to the correct result while still allowing the work done in each expansion to be calculated. Figure A-4 shows divisions into six equal volume increments, $\Delta V''$ with a resulting total work

$$w = P_1\Delta V'' + P_2\Delta V'' + \cdots$$
$$= \sum_{i=1}^{6} P_i\Delta V''$$

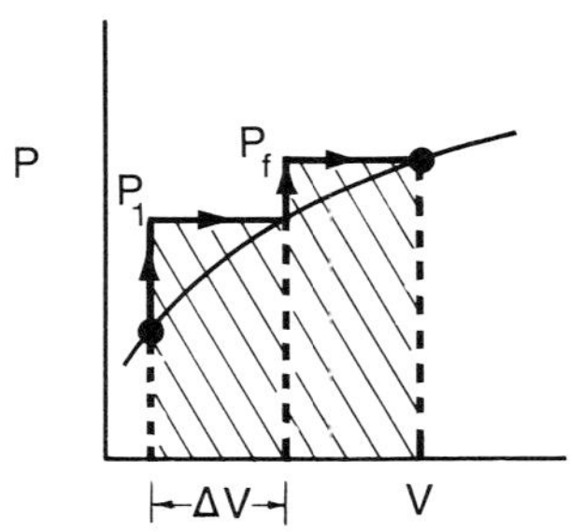

Figure A-3

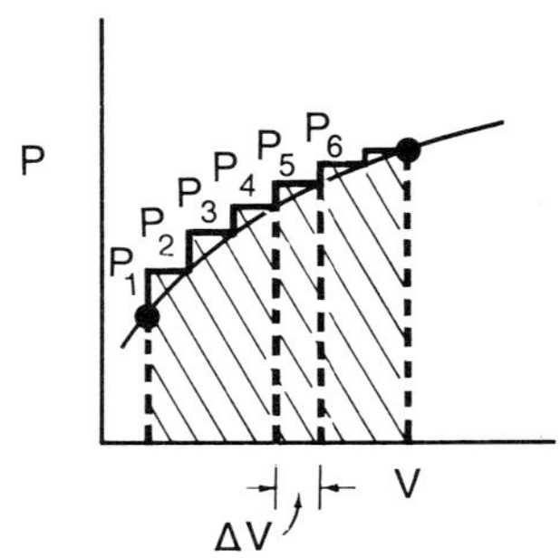

Figure A-4

The Greek letter sigma, Σ, indicates a summation of the products $P_i\Delta V''$ for each value of i. In our particular case i can have the values 1, 2, 3 . . . 6 as shown above and below the sigma sign.

Our aim is now clear. If we continue reducing the size of ΔV, in the limit that ΔV becomes infinitesimally small, we will have a path that is indistinguishable from the desired one while allowing us to calculate the work done during each tiny expansion. This problem is dealt with directly in integral calculus. An infinitesimal change in volume is given the notation dV and the sum of the areas of each narrow rectangle which will give us the work done is indicated by

$$w = \int_{V_0}^{V_f} PdV$$

called the "integral of PdV between the limits V_0 and V_f." The integral sign is related to the letter "*S*," implying a summation of all the little areas PdV.

Two cases of particular interest in thermodynamics permit a simpler solution of this equation.

(1) Pressure constant: If the pressure is constant it can be removed outside the integral sign

$$w = P\int_{V_0}^{V_f} dV$$

with the simple result, already familar to us, that

$$w = P(V_f - V_o) = P\Delta V$$

(2) Pressure not constant; ideal gas expanded at equilibrium pressure: if the gas used is ideal (all gases are ideal at low pressures), we can carry out the expansion so that the pressure is inversely related to the volume at every instant,

$$P = \frac{nRT}{V}$$

Substitution in the integral gives us

$$w = nRT \int_{V_o}^{V_f} \frac{dV}{V}$$

The result, stated here without proof, is derived in every introductory calculus class to be

$$w = nRT(\log_e V_f - \log_e V_o)$$
$$w = nRT \log_e\left(\frac{V_f}{V_o}\right)$$

We may conclude by generalizing these two results to any function y of a variable x

$$y = c \text{ (constant)} \qquad \int_{x_1}^{x_2} c\,dx = c\int_{x_1}^{x_2} dx = c(x_2 - x_1) = c\Delta x$$

$$y = \frac{1}{x} \qquad \int_{x_1}^{x_2} \frac{dx}{x} = \log_e x_2 - \log_e x_1 = \log_e\left(\frac{x_2}{x_1}\right)$$

A brief description of the nature of $\log_e$ will now be given.

The exponential e

We just noted that the integral of a commonly occurring function $1/x$ is given by the logarithm to the base "e" of x. What is "e?" This number arises from further consideration of the integral calculus. Any number y can be expressed as some other number a raised to a power, x.

$$y = a^x$$

There is one particular value of a that has great importance because it has the unique property of making y the integral of itself. This particular value of a is called e.

$$\int y\,dx = \int e^x dx = e^x = y$$

in which

$$e = 2.7182.\ .\ .\ .\ .$$

In general, if

$$y = e^x$$

then the number x is called the logarithm of y to the base e:

$$x = \log_e y$$

or, in a usual notation

$$x = \ln y$$

The symbol ln is used to represent the natural logarithm, or the logarithm to the base e. This logarithm is simply related to the more common logarithm to the base 10:

$$\log_{10} y = 0.4343 \log_e y$$
$$\log_e y = 2.303 \log_{10} y$$

When the function $y = e^x$ is plotted against x, the solid curve shown in Figure A-5 results. "Exponential" curves such as these are frequently encountered in chemistry and physics, especially when some form of growth is being considered. This is so because the slope, or rate of increase of y, is always proportional to the value of y itself. Thus the curve is a regularly increasing one. As an example we might expect population growth to follow an exponential rise since the number of births probably increases in the same proportion as the population itself increases.

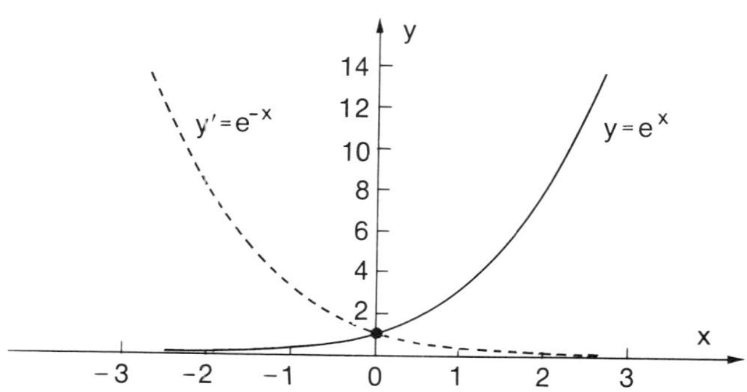

Figure A-5

Of course, the function e^{-x} is simply related to the e^x curve. If $y = e^x$, then $1/y = e^{-x}$. So the value of e^{-x} for any particular value of x is just the reciprocal of e^{+x}. The dashed curve in Figure A-5 shows a plot of $y' = e^{-x}$.

appendix b
thermodynamic properties

Formula	State	ΔH_f^o (kcal/mole)	ΔG_f^o (kcal/mole)	S^o (cal/mole degK)	C_p^o (cal/mole degK)
		at 298.16°K (25°C)			
Aluminum					
Al	s	0.0	0.0	6.8	5.8
Al_2O_3	s (α, corundum)	−400.5	−378.2	12.2	18.9
AlF_3	s	−359.5	−340.6	15.9	18.0
$AlCl_3$	s	−168.3	−150.3	26.5	22.0
$AlBr_3$	s	−126.0	−122.1	49.0	24.5
AlI_3	s	−75.0	−71.9	38.0	23.6
$Al_2(SO_4)_3$	s	−822.4	−741.0	57.2	62.0
AlN	s	−76.0	−68.6	4.82	7.20
Antimony					
Sb	s	0.0	0.0	10.9	6.0
Sb_2O_3	s	−168.4	−149.0	29.4	24.2
Sb_2O_5	s	−232.3	−198.2	29.9	28.1
$SbCl_3$	g	−75.0	−72.0	80.7	18.3
	s	−91.3	−77.4	44.0	25.8
Sb_2S_3	s (black)	−41.8	−41.5	43.5	28.6
Argon					
Ar	g	0.0	0.0	37.0	5.0
Arsenic					
As	s (α grey)	0.0	0.0	8.4	5.9
	g	72.3	62.4	41.6	5.0
As_4	g	34.4	22.1	75.0	–
As_4O_6	s (octahedral)	−314.0	−275.5	51.2	45.7
As_2O_5	s	−221.0	−187.0	25.2	27.8
AsH_3	g	15.9	16.5	53.2	9.1
As_2S_3	s	−40.4	−40.3	39.1	27.8
Barium					
Ba	s	0.0	0.0	16.0	6.3
BaO	s	−133.4	−126.3	16.8	11.3
BaO_2	s	−150.5	–	–	–
$BaCl_2$	s	−205.6	−193.8	30.0	18.0
$BaCl_2 \cdot H_2O$	s	−278.4	−253.1	40.0	28.2
$BaCl_2 \cdot 2H_2O$	s	−349.3	−309.7	48.5	37.1
$BaSO_4$	s	−350.2	−323.4	31.6	24.3
$Ba(NO_3)_2$	s	−237.1	−190.1	51.1	36.1
$BaCO_3$	s (witherite)	−291.3	−272.2	26.8	20.4
Beryllium					
Be	s	0.0	0.0	2.3	4.3
BeO	s	−146.0	−139.0	3.4	6.1
$BeCl_2$	s	−112.0	−101.9	21.5	–

		ΔH_f^o	ΔG_f^o	S^o	C_p^o
Bismuth					
Bi	s	0.0	0.0	13.6	6.1
Bi_2O_3	s	−137.9	−118.7	36.2	27.2
$BiCl_3$	g	−64.7	−62.2	85.3	19.0
	s	−90.6	−76.2	45.3	–
BiOCl	s	−87.3	−77.0	20.6	–
Bi_2S_3	s	−43.8	−39.4	35.3	30.7
Boron					
B	s	0.0	0.0	1.4	2.6
B_2O_3	s	−304.2	−285.3	12.90	15.0
B_2H_6	g	7.5	19.8	55.7	13.5
BF_3	g	−265.4	−261.3	60.7	12.1
BCl_3	g	−96.5	−92.9	69.3	15.0
	ℓ	−102.1	−92.6	49.3	25.5
BBr_3	g	−44.6	−51.0	77.5	16.3
Bromine					
Br_2	ℓ	0.0	0.0	36.4	18.1
	g	7.4	0.7	58.6	8.6
Br	g	26.7	19.7	41.8	5.0
Cadmium					
Cd	s (α)	−0.1	−0.1	12.4	6.2
CdO	s	−61.7	−54.6	13.1	10.4
$CdCl_2$	s	−93.6	−82.2	27.5	17.9
$CdCl_2 \cdot H_2O$	s	−164.5	−140.3	40.1	–
$CdCl_2 \cdot 2.5H_2O$	s	−270.5	−225.6	54.3	–
CdS	s	−38.7	−37.4	15.5	13.2
$CdSO_4$	s	−223.1	−196.6	29.4	23.8
Cesium					
Cs	s	0.0	0.0	19.8	7.4
Cs_2O	s	−75.9	–	–	–
CsH	g	29.0	24.3	51.2	7.1
CsF	s	−126.9	−119.7	19.8	12.1
CsCl	s	−103.5	−96.8	23.9	12.6
CsBr	s	−94.3	−91.6	29.0	12.4
CsI	s	−80.5	−79.7	31.0	12.4
Calcium					
Ca	s	0.0	0.0	9.9	6.3
CaO	s	−151.9	−144.4	9.5	10.2
$Ca(OH)_2$	s	−235.8	−214.3	18.2	20.2
$CaCl_2$	s	−190.0	−179.3	27.2	17.4
$CaSO_4$	s (anhydrite)	−342.4	−315.6	25.5	23.8
$CaSO_4 \cdot 0.5H_2O$	s (α)	−376.5	−343.0	31.2	28.6
	s (β)	−376.0	−342.8	32.1	29.6
$CaSO_4 \cdot 2H_2O$	s	−483.1	−429.2	46.4	44.5
CaC_2	s	−15.0	−16.2	16.8	14.9
$CaCO_3$	s (calcite)	−288.4	−269.8	22.2	19.6

		ΔH_f°	ΔG_f°	S°	C_p°
Carbon					
C	s (graphite)	0.0	0.0	1.37	2.0
	s (diamond)	0.45	0.69	0.57	1.5
	g	171.3	160.4	37.8	5.0
CO	g	−26.4	−32.8	47.2	7.0
CO_2	g	−94.1	−94.3	51.1	8.9
CH_4	g	−17.9	−12.1	44.5	8.4
C_2H_6	g	−20.2	−7.9	54.8	12.6
C_2H_4	g	12.5	16.3	52.5	10.4
C_2H_2	g	54.2	50.0	48.0	10.5
C_3H_8	g	−24.8	−5.6	64.5	17.6
C_6H_6	g	19.8	31.0	64.3	19.5
	ℓ	11.7	29.8	41.3	—
HCHO	g	−27.7	−27.0	52.3	8.5
HCOOH	g	−90.5	−80.2	60.0	—
CH_3OH	g	−48.0	38.7	57.3	10.5
	ℓ	−57.0	−39.8	30.3	19.5
CCl_4	g	−24.6	14.5	74.0	19.9
	ℓ	−32.4	−15.6	51.7	31.5
$COCl_2$	g	−52.3	−48.9	67.7	13.8
CH_3Cl	g	−19.3	−13.7	56.0	9.7
CH_2Cl_2	g	−22.1	−15.8	64.6	12.2
$CHCl_3$	g	−24.6	−16.8	70.6	15.7
	ℓ	−32.1	−17.6	48.2	27.2
CS_2	g	28.0	16.0	56.8	10.8
	ℓ	21.4	15.6	36.2	18.1
Chlorine					
Cl_2	g	0.0	0.0	53.3	8.1
Cl	g	29.1	25.3	39.5	5.2
Chromium					
Cr	s	0.0	0.0	5.7	5.6
Cr_2O_3	s	−269.7	−250.2	19.4	28.4
CrO_3	s	−138.4	−119.9	17.2	—
$CrCl_2$	s	−94.6	−85.1	27.4	16.9
$CrCl_3$	s	−134.6	−118.0	30.0	21.5
Cobalt					
Co	s	0.0	0.0	7.2	6.1
CoO	s	−57.2	−51.5	12.6	—
Co_3O_4	s	−216.3	−188.0	24.5	—
$CoCl_2$	s	−77.8	−67.5	25.4	18.8
Copper					
Cu	s	0.0	0.0	8.0	5.8
Cu_2O	s	−39.8	−34.4	22.4	16.7
CuO	s	−37.1	−30.4	10.2	10.6
CuCl	s	−32.2	−28.4	21.9	11.6
$CuCl_2$	s	−49.2	−39.0	27.0	19.0
Cu_2S	s	−19.0	−20.6	28.9	18.2
CuS	s	−11.6	−11.7	15.9	11.4
$CuSO_4$	s	−184.0	−158.2	27.1	24.1
$CuSO_4 \cdot H_2O$	s	−259.0	−219.2	35.8	31.3

		ΔH_f^o	ΔG_f^o	S^o	C_p^o
Copper—*cont.*					
$CuSO_4 \cdot 3H_2O$	s	−402.3	−334.6	53.8	49.0
$CuSO_4 \cdot 5H_2O$	s	−544.5	−449.3	73.0	67.2
$CuCO_3$	s	−142.2	−123.8	21.0	—
Fluorine					
F_2	g	0.0	0.0	48.4	7.5
F	g	18.9	14.8	37.9	5.4
F_2O	g	5.5	9.7	59.0	—
Germanium					
Ge	s	0.0	0.0	7.4	5.6
GeO_2	amorphous	−128.4	−114.4	12.5	—
$GeCl_4$	ℓ	−127.1	−110.6	58.7	—
Gold					
Au	s	0.0	0.0	11.4	6.0
Au_2O_3	s	19.3	39.0	30.0	—
$Au(OH)_3$	s	−100.0	−69.4	29.0	—
$AuCl_3$	s	−28.3	−11.6	35.4	—
Helium					
He	g	0.0	0.0	30.1	5.0
Hydrogen					
H_2	g	0.0	0.0	31.2	6.9
H_2O	g	−57.8	−54.6	45.1	8.0
	ℓ	−68.3	−56.7	16.7	18.0
H_2O_2	g	−32.6	−25.2	55.6	10.3
HF	g	−64.8	−65.3	41.5	7.0
HCl	g	−22.1	−22.8	44.6	7.0
HBr	g	−8.7	−12.8	47.5	7.0
HI	g	6.3	0.4	49.3	7.0
H_2S	g	−4.9	−8.0	49.2	8.2
H_2Se	g	7.1	3.8	52.3	8.3
H_2Te	g	36.9	33.1	56.0	—
Iodine					
I_2	s	0.0	0.0	27.8	13.0
	g	14.9	4.6	62.3	8.8
I	g	25.5	16.8	43.2	5.0
ICl	g	4.2	−1.3	59.1	8.5
ICl_3	s	−21.1	−5.4	41.1	—
Iron					
Fe	s	0.0	0.0	6.5	6.0
$Fe_{0.95}O$	s (wustite)	−63.2	−57.5	14.0	11.5
Fe_2O_3	s (hematite)	−196.5	−177.1	21.5	25.0
Fe_3O_4	s (magnetite)	−267.0	−242.8	36.2	36.3

		ΔH_f^o	ΔG_f^o	S^o	C_p^o
Iron—*cont.*					
$FeCl_2$	s	−81.5	−72.2	28.6	18.2
$FeCl_3$	s	−96.8	−80.6	32.2	22.8
FeS	s (α)	−22.7	−23.3	16.1	13.1
FeS_2	s (pyrites)	−42.5	−39.8	12.7	14.8
$FeSO_4$	s	−220.5	−194.8	25.7	–
Fe_3C	s (cementite)	5.0	3.5	25.7	25.3
$FeCO_3$	s (siderite)	−178.7	−161.1	22.2	19.6
Krypton					
Kr	g	0.0	0.0	39.2	5.0
Lead					
Pb	s	0.0	0.0	15.5	6.3
PbO	s (red)	−52.3	−45.2	15.9	11.0
	s (yellow)	−51.9	−44.9	16.4	10.9
PbO_2	s	−66.3	−51.9	16.4	15.4
Pb_3O_4	s	−171.7	−143.7	50.5	35.1
PbF_2	s	−158.7	−147.5	26.4	–
$PbCl_2$	s	−85.9	−75.1	32.5	–
$PbBr_2$	s	−66.6	−62.6	38.6	19.2
PbI_2	s	−41.9	−41.5	41.8	18.5
PbS	s	−24.0	−23.6	21.8	11.8
$PbSO_4$	s	−219.9	−194.4	35.5	24.7
Lithium					
Li	s	0.0	0.0	6.7	5.6
	g	37.1	29.2	33.1	5.0
Li_2O	s	−142.4	−133.8	9.1	13.0
LiH	g	30.7	25.2	40.8	7.1
LiOH	s	−116.4	−105.5	10.2	11.9
LiF	s	−146.3	−139.6	8.6	10.0
LiCl	s	−97.7	−91.9	13.9	12.0
Li_2SO_4	s	−342.8	−316.0	29.0	–
Li_2CO_3	s	−290.5	−270.7	21.6	23.3
Magnesium					
Mg	s	0.0	0.0	7.8	5.7
MgO	s	−143.8	−136.1	6.4	8.9
$Mg(OH)_2$	s	−221.0	−199.3	15.1	18.4
MgF_2	s	−263.5	−250.8	13.7	14.7
$MgCl_2$	s	−153.4	−141.6	21.4	17.0
$MgCl_2 \cdot H_2O$	s	−231.1	−206.1	32.8	27.5
$MgCl_2 \cdot 2H_2O$	s	−306.0	−267.3	43.0	38.0
$MgCl_2 \cdot 4H_2O$	s	−454.0	−390.5	63.1	57.7
$MgCl_2 \cdot 6H_2O$	s	−597.4	−505.6	87.5	75.5
$MgSO_4$	s	−305.5	−280.5	21.9	23.0
$Mg(NO_3)_2$	s	−188.7	−140.6	39.2	33.9
$MgCO_3$	s	−266.0	−246.0	15.7	18.0

		ΔH_f°	ΔG_f°	S°	C_p°
Manganese					
Mn	s (α)	0.0	0.0	7.6	6.3
MnO	s	−92.0	−86.8	14.4	10.3
MnO_2	s	−124.5	−111.4	12.7	12.9
Mn_2O_3	s	−229.4	−210.8	26.4	25.8
Mn_3O_4	s	−331.4	−306.0	35.5	33.3
$Mn(OH)_2$	amorphous	−165.8	−145.9	21.1	—
MnF_2	s	−189.0	−179.0	22.2	16.2
$MnCl_2$	s	−112.0	−102.2	28.0	17.4
$MnBr_2$	s	−90.7	−87.4	33.0	—
MnI_2	s	−59.3	−59.5	36.0	—
$MnSO_4$	s	−254.2	−228.5	26.8	23.9
$MnCO_3$	s	−213.9	−195.4	20.5	19.5
Mercury					
Hg	ℓ	0.0	0.0	18.2	6.6
	g	14.5	7.6	41.8	5.0
HgO	s (red)	−21.7	−14.0	16.8	10.9
	s (yellow)	−21.6	−14.1	17.5	—
Hg_2Cl_2	s	−63.3	−50.3	46.8	24.3
$HgCl_2$	s	−55.0	−44.0	34.5	18.3
HgS	s (red)	−13.9	−11.7	18.6	11.1
	s (black)	−12.9	−11.1	19.9	11.1
Hg_2SO_4	s	−177.3	−149.1	48.0	31.5
Molybdenum					
Mo	s	0.0	0.0	6.8	5.6
MoO_2	s	−139.5	−126.9	13.6	—
MoO_3	s	−180.3	−161.9	18.7	17.6
MoS_2	s	−55.5	−53.8	15.1	15.2
Mo_2S_3	s	−102.0	−99.5	28.0	—
Neon					
Ne	g	0.0	0.0	34.9	5.0
Nickel					
Ni	s	0.0	0.0	7.2	6.2
NiO	s	−58.4	−51.7	9.2	10.6
$NiCl_2$	s	−73.0	−61.9	23.3	18.6
$NiCl_2 \cdot 6H_2O$	s	−505.8	−410.5	75.2	—
$NiSO_4$	s	−213.0	−184.9	18.6	33.4
$NiSO_4 \cdot 6H_2O$	s (blue)	−642.5	−531.0	73.1	82.0
$Ni(CO)_4$	g	−144.7	−140.3	96.0	—
Nitrogen					
N_2	g	0.0	0.0	45.8	7.0
NO	g	21.6	20.7	50.3	7.1
NO_2	g	7.9	12.3	57.3	8.9
N_2O	g	19.6	24.9	52.5	9.2
N_2O_4	g	2.2	23.4	72.7	18.5
N_2O_5	g	2.7	27.5	85.0	20.2
NH_3	g	−11.0	−3.9	46.0	8.4
HNO_3	ℓ	−41.6	−19.3	37.2	26.3
NH_4NO_3	s	−87.4	−44.0	36.1	33.3

		ΔH_f^o	ΔG_f^o	S^o	C_p^o
Nitrogen—*cont.*					
NH_4Cl	s	−75.2	−48.5	22.6	20.1
$(NH_4)_2SO_4$	s	−282.2	−215.6	52.6	44.8
Oxygen					
O_2	g	0.0	0.0	49.0	7.0
O_3	g	34.1	39.0	57.1	9.4
Phosphorus					
P	s (white)	0.0	0.0	9.8	5.7
	s (red)	−4.2	−2.9	5.4	5.1
	g	75.2	66.5	39.0	5.0
P_4	g	14.1	5.8	66.9	16.0
P_4O_{10}	s	−713.2	−644.8	54.7	50.6
PH_3	g	1.3	3.2	50.2	8.9
H_3PO_4	s	−305.7	−267.5	26.4	25.4
PCl_3	g	−68.6	−64.0	74.5	17.2
PCl_5	g	−89.6	−72.9	87.1	27.0
$POCl_3$	g	−133.5	−122.6	77.8	20.3
PBr_3	g	−33.3	−38.9	83.2	18.2
Platinum					
Pt	s	0.0	0.0	10.0	6.3
$Pt(OH)_2$	s	−87.2	−68.2	26.5	—
Potassium					
K	s	0.0	0.0	15.2	7.0
K_2O	s	−86.4	−77.0	23.5	—
KH	g	30.0	25.1	47.3	—
KF	s	−134.5	−127.4	15.9	11.7
KCl	s	−104.2	−97.6	19.8	12.3
$KClO_3$	s	−93.5	−69.3	34.2	24.0
KBr	s	−93.7	−90.6	23.0	12.8
$KBrO_3$	s	−79.4	−58.2	35.6	25.1
KI	s	−78.3	−77.0	24.9	13.2
KIO_3	s	−121.5	−101.7	36.2	25.4
K_2SO_4	s	−342.7	−314.6	42.0	31.1
KNO_3	s	−117.8	−94.0	31.8	23.0
$KMnO_4$	s	−194.4	−170.6	41.0	28.5
KOH	s	−101.8	—	—	—
Radon					
Rn	g	0.0	0.0	42.1	5.0
Rubidium					
Rb	s	0.0	0.0	18.3	7.3
RbF	s	−131.3	−124.0	18.0	12.6
RbCl	s	−102.9	−96.0	21.9	12.3
RbBr	s	−93.0	−90.4	25.9	12.7
RbI	s	−78.5	−77.8	28.2	12.5
Selenium					
Se	s (grey)	0.0	0.0	10.1	6.1
	g	54.3	44.7	42.2	5.0

		ΔH_f^0	ΔG_f^0	S^0	C_p^0
Selenium—*cont.*					
Se_2	g	34.9	23.0	60.2	8.5
SeF_6	g	−267.0	−243.0	75.0	26.4
Silicon					
Si	s	0.0	0.0	4.5	4.8
SiO_2	s (quartz)	−217.7	−204.8	10.0	10.6
SiH_4	g	8.2	13.6	48.9	10.2
SiF_4	g	−386.0	−375.9	67.5	17.6
$SiCl_4$	g	−157.0	−147.5	79.0	21.6
SiC	s (β cubic)	−15.6	−15.0	4.0	6.4
	s (α hexagonal)	−15.0	−14.4	3.9	6.4
Silver					
Ag	s	0.0	0.0	10.2	6.1
Ag_2O	s	−7.3	−2.6	29.1	15.7
AgF	s	−48.5	−44.2	20.0	–
AgCl	s	−30.4	−26.2	23.0	12.1
AgBr	s	−23.8	−22.9	25.6	12.5
AgI	s	−14.9	−15.8	27.3	13.0
Ag_2S	s (rhombic α)	−7.6	−9.6	34.8	18.0
Ag_2SO_4	s	−170.5	−147.2	47.8	31.4
$AgNO_3$	s	−29.4	−7.7	33.7	22.2
Ag_2CO_3	s	−120.9	−104.5	40.0	26.8
Sodium					
Na	s	0.0	0.0	12.2	6.8
Na_2O	s	−99.4	−90.0	17.4	16.3
Na_2O_2	s	−123.0	−107.8	22.6	–
NaH	g	29.9	24.8	44.9	7.0
NaOH	s	−102.0	−91.0	15.3	14.3
NaF	s	−136.5	−129.3	12.3	11.0
NaCl	s	−98.6	−92.2	17.4	11.9
NaBr	s	−86.0	−82.9	20.0	12.5
NaI	s	−68.8	−67.5	21.8	–
Na_2SO_4	s	−330.9	−302.8	35.7	30.5
$Na_2SO_4 \cdot 10H_2O$	s	−1033.5	−870.9	141.7	140.4
$NaNO_3$	s	−111.5	−87.5	27.8	22.2
Na_2CO_3	s	−270.3	−250.4	32.5	26.4
$Na_2CO_3 \cdot 10H_2O$	s	−975.6	–	–	128.0
$NaHCO_3$	s	−226.5	−203.6	24.4	20.9
Strontium					
Sr	s	0.0	0.0	13.0	6.0
SrO	s	−141.1	−133.8	13.0	10.8
$SrCl_2$	s	−198.0	−186.7	28.0	18.9
$SrSO_4$	s	−345.3	−318.9	29.1	–
$SrCO_3$	s	−291.2	−271.9	23.2	19.5
$Sr(OH)_2$	s	−229.3	–	–	–

		ΔH_f°	ΔG_f°	S°	C_p°
Strontium – *cont.*					
SrF_2	s	−171.1	–	–	19.0
SrI_2	s	−135.5	–	–	19.5
Sulfur					
S	s (rhombic)	0.0	0.0	7.6	5.4
	s (mono-clinic)	0.08	–	–	–
	g	66.6	56.9	40.1	5.7
S_8	g	24.4	11.9	103.0	37.4
SO_2	g	−70.9	−71.7	59.3	9.5
SO_3	g	−94.6	−88.7	61.3	12.1
H_2SO_4	ℓ	−194.5	−164.9	37.5	33.2
SF_6	g	−289.0	−264.2	69.7	23.2
Tellurium					
Te	s	0.0	0.0	11.9	6.1
TeO_2	s	−77.1	−64.6	19.0	–
TeF_6	g	−315.0	−292.0	80.7	–
Tin					
Sn	s (white)	0.0	0.0	12.3	6.5
	s (grey)	−0.5	0.0	10.5	6.2
SnO	s	−68.3	−61.4	13.5	10.6
SnO_2	s	−138.8	−124.2	12.5	12.6
$SnCl_4$	ℓ	−122.2	−105.2	61.8	39.5
SnS	s	−24.0	−23.5	18.4	11.8
SnS_2	s	−40.0	−38.0	20.9	16.8
Titanium					
Ti	s	0.0	0.0	7.2	6.0
TiO_2	s (rutile)	−218.0	−203.8	12.0	13.2
$TiCl_4$	ℓ	−191.5	−175.3	59.6	37.5
TiN	s	−73.0	−66.1	7.2	8.9
TiC	s	−54.0	−53.0	5.8	8.0
Tungsten					
W	s	0.0	0.0	8.0	6.0
WO_3	s (yellow)	−200.8	−182.5	19.9	19.5
Uranium					
U	s	0.0	0.0	12.0	6.6
UO_2	s	−270.0	−257.0	18.6	15.3
UF_6	g	−505.0	−485.0	90.8	–
Vanadium					
V	s	0.0	0.0	7.0	5.8
V_2O_3	s	−290.0	−271.0	23.6	24.8
VO_2	s	−172.0	−159.0	12.3	14.2
V_2O_5	s	−373.0	−344.0	31.3	31.0
VCl_2	s	−108.0	−97.0	23.2	17.3
VCl_3	s	−137.0	−120.0	31.3	22.3

		ΔH_f^o	ΔG_f^o	S^o	C_p^o
Xenon					
Xe	g	0.0	0.0	40.5	5.0
Zinc					
Zn	s	0.0	0.0	10.0	6.1
ZnO	s	−83.2	−76.1	10.4	9.6
$Zn(OH)_2$	s	−153.5	—	—	17.3
$ZnCl_2$	s	−99.2	−88.3	26.6	17.0
$ZnBr_2$	s	−78.5	−74.6	33.1	—
ZnI_2	s	−49.7	−49.9	38.5	—
ZnS	s (zinc blende)	−49.2	−48.1	13.8	11.0
	s (wurtzite)	−46.0	—	—	—
$ZnSO_4$	s	−234.9	−209.0	28.6	28.0
$ZnSO_4 \cdot H_2O$	s	−311.8	−270.6	33.1	34.7
$ZnSO_4 \cdot 6H_2O$	s	−663.8	−555.6	86.9	85.5
$ZnSO_4 \cdot 7H_2O$	s	−735.6	−612.6	92.9	91.6
$ZnCO_3$	s	−194.3	−174.9	19.7	19.1
Zirconium					
Zr	s	0.0	0.0	9.2	6.2
ZrO_2	s	−258.2	−244.4	12.0	13.4
$ZrCl_4$	s	−234.7	−213.4	44.5	28.7

index